In Situ Hybridization

The Practical Approach Series

SERIES EDITOR

B. D. HAMES
Department of Biochemistry and Molecular Biology
University of Leeds, Leeds LS2 9JT, UK

See also the Practical Approach web site at **http://www.oup.co.uk/PAS**

★ **indicates new and forthcoming titles**

Affinity Chromatography
Affinity Separations
Anaerobic Microbiology
Animal Cell Culture (2nd edition)
Animal Virus Pathogenesis
Antibodies I and II
Antibody Engineering
★ Antisense Technology
Applied Microbial Physiology
Basic Cell Culture
Behavioural Neuroscience
Bioenergetics
Biological Data Analysis
★ Biomaterial Immobilization
Biomechanics – Materials
Biomechanics – Structures and Systems
Biosensors
Carbohydrate Analysis (2nd edition)
Cell–Cell Interactions
The Cell Cycle
Cell Growth and Apoptosis

★ Cell Separation
Cellular Calcium
Cellular Interactions in Development
Cellular Neurobiology
★ Chromatin
Clinical Immunology
★ Complement
Crystallization of Nucleic Acids and Proteins
Cytokines (2nd edition)
The Cytoskeleton
Diagnostic Molecular Pathology I and II
DNA and Protein Sequence Analysis
DNA Cloning 1: Core Techniques (2nd edition)
DNA Cloning 2: Expression Systems (2nd edition)
DNA Cloning 3: Complex Genomes (2nd edition)
DNA Cloning 4: Mammalian Systems (2nd edition)

In Situ Hybridization

A Practical Approach

Second Edition

Edited by

D. G. WILKINSON

NIMR, London

Oxford   New York   Tokyo
OXFORD UNIVERSITY PRESS
1998

Oxford University Press, Great Clarendon Street, Oxford OX2 6DP

Oxford   New York

Athens   Auckland   Bangkok   Bogota   Bombay   Buenos Aires   Calcutta
Cape Town   Dar es Salaam   Delhi   Florence   Hong Kong   Istanbul
Karachi   Kuala Lumpur   Madrid   Melbourne   Mexico City   Mumbai
Nairobi   Paris   São Paolo   Singapore   Taipei   Tokyo   Toronto   Warsaw

and associated companies in
Berlin   Ibadan

Oxford is a trade mark of Oxford University Press

Published in the United States
by Oxford University Press Inc., New York

Users of books in the Practical Approach Series are advised that prudent
laboratory safety procedures should be followed at all times. Oxford
University Press makes no representation, express or implied, in respect of
the accuracy of the material set forth in books in this series and cannot
accept any legal responsibility or liability for any errors or omissions
that may be made.

A catalogue record for this book is available from the British Library

Library of Congress Cataloging in Publication Data
Wilkinson, D. G.
In situ hybridization: a practical approach/edited by D. G.
Wilkinson
(The practical approach series; 196)
Includes bibliographical references and index.
1. In situ hybridization–Laboratory manuals.   I. Series.
QH452.8.I52   1998   572.8'4–dc21   98–7769

ISBN 0 19 963659 1 (Hbk)
0 19 963658 3 (Pbk)

Typeset by Footnote Graphics, Warminster, Wilts
Printed in Great Britain by Information Press, Ltd, Eynsham, Oxon.

Preface

Visualization of the location of genes on chromosomes or of specific mRNAs or viruses in tissues is crucial for elucidation of the structure, regulation, and function of genes. The technique of *in situ* hybridization, which reveals the location of specific nucleic acid sequences, has therefore been invaluable for both fundamental and applied research, including studies of gene function during embryo development, and analysis of chromosome rearrangements associated with genetic disorders.

A number of important technical advances have been made since the first edition of this book, in particular in refining procedures for the whole mount detection of mRNAs and for the detection of multiple genes; the establishment of non-radioactive methods for simultaneous detection of several mRNAs; and the use of PCR-based techniques for the amplification of signal. These latter PCR techniques are described in another volume in this series: PCR *in situ* hybridization.

In this volume, detailed step-by-step protocols are presented for the major variations of the technique of *in situ* hybridization. Chapter 1 provides an overview of theoretical and practical aspects of the important steps of *in situ* hybridization. The next four chapters describe techniques for visualizing a specific mRNA at single cell (or slightly lower) resolution. Chapters 2 and 3 describe protocols for using oligonucleotide probes (radioactive and hapten labelled) and radiolabelled RNA probes, respectively, on tissue sections. Chapter 4 describes the use of hapten labelled RNA probes for sections and whole mount tissues. Chapter 5 describes various techniques for detecting multiple mRNAs and for combining *in situ* hybridization with immunocytochemistry. The higher resolution detection of mRNAs and viruses at the subcellular level using the electron microscope is described in Chapter 6. Chapter 7 describes techniques for locating specific sequences on chromosomes by using fluorescence detection methods. Finally, Chapter 8 discusses a crucial new development for the studies of gene regulation and function: the establishment of gene expression databases.

All of these chapters are written by authors with extensive practical experience of establishing reliable techniques of *in situ* hybridization, and the procedures are widely applicable to many systems. I hope that they will prove useful to all who wish to learn or optimize the procedures of *in situ* hybridization.

London   D. G. W.
November 1997

Contents

Contents

7. Detection of genomic sequences by
  fluorescence *in situ* hybridization to
  chromosomes 161

Lyndal Kearney

8. Gene expression databases   190

Duncan Davidson, Richard Baldock, Jonathan Bard,
Matthew Kaufman, Joel E. Richardson, Janan T. Eppig,
and Martin Ringwald

Contents

Contributors

RICHARD BALDOCK
MRC Human Genetics Unit, Western General Hospital, Crewe Road, Edinburgh EH4 2XU, UK.

JONATHAN BARD
Anatomy Department, University of Edinburgh, Teviot Place, Edinburgh, UK.

DUNCAN DAVIDSON
MRC Human Genetics Unit, Western General Hospital, Crewe Road, Edinburgh EH4 2XU, UK.

JANAN T. EPPIG
The Jackson Laboratory, 600 Main Street, Bar Harbor, ME 04609, USA.

T. JOWETT
Department of Biochemistry and Genetics, The Medical School, The University, Newcastle upon Tyne NE2 4HH, UK.

MATTHEW KAUFMAN
Anatomy Department, University of Edinburgh, Teviot Place, Edinburgh, UK.

LYNDAL KEARNEY
MRC Molecular Haematology Unit, Institute of Molecular Medicine, John Radcliffe Hospital, Headington, Oxford OX3 9DU, UK.

GREGORY J. MICHAEL
Department of Anatomy, Basic Medical Sciences, Queen Mary and Westfield College, Mile End Road, London E1 4NS, UK.

FRANCINE PUVION-DUTILLEUL
Laboratoire Organisation fonctionnelle du Noyau de l'UPR 9044 CNRS, 7 rue Guy Môquet, BP No. 8, F-94801 Villejuif Cedex, France.

MARCUS RATTRAY
Division of Biochemistry and Molecular Biology, UMDS, Guy's Hospital, St Thomas Street, London SE1 9RT, UK.

JOEL E. RICHARDSON
The Jackson Laboratory, 600 Main Street, Bar Harbor, ME 04609, USA.

MARTIN RINGWALD
The Jackson Laboratory, 600 Main Street, Bar Harbor, ME 04609, USA.

Contributors

ANTONIO SIMEONE
International Institute of Genetics and Biophysics, CNR, Via G. Marconi, 12, 80125 Naples, Italy.

DAVID G. WILKINSON
Division of Developmental Neurobiology, National Institute for Medical Research, The Ridgeway, Mill Hill, London NW7 1AA, UK.

QILING XU
Division of Developmental Neurobiology, National Institute for Medical Research, The Ridgeway, Mill Hill, London NW7 1AA, UK.

Abbreviations

A	adenine (or adenosine)
A_{260}	absorbance at 260 nm wavelength
ABC	avidin–biotinylated enzyme complex
AMCA	aminomethylcoumarin acetic acid
AP	alkaline phosphatase
ATP	adenosine triphosphate
BAC	bacterial artificial chromosome
BCIP	5-bromo-4-chloro-3-indolyl phosphate
Bio	biotin
BrdU	5-bromodeoxyuridine
BSA	bovine serum albumin
C	cytosine (or cytidine)
CCD	charge-coupled device
cDNA	complementary DNA
CDS	coding region
CGH	comparative genomic hybridization
CHAPS	3-[(3-cholamidopropyl)dimethylammonio]-1-propane sulfonate
Ci	Curies
CLL	chronic lymphocytic leukaemia
CTP	cytidine triphosphate
DAB	diaminobenzidine
DAPI	4′,6-diamidino-2-phenylindole
dATP	deoxyadenosine triphosphate
DEPC	diethylpyrocarbonate
DIC	differential interference contrast
DIG	digoxigenin
DMSO	dimethyl sulfoxide
DNase	deoxyribonuclease
dsDNA	double-stranded DNA
DTT	dithiothreitol
EDTA	ethylenediaminetetraacetic acid
ETS	external transcribed spacer
FCS	fetal calf serum
FISH	fluorescence *in situ* hybridization
FITC	fluorescein isothiocyanate
FLU	fluorescein
G	guanine (or guanosine)
GTP	guanosine triphosphate
GXD	Gene Expression Database
HFV	human foamy virus

HSV	herpes simplex virus
K_d	equilibrium constant
IgG	immunoglobulin G
ITS	internal transcribed spacer
KTBT	Tris-buffered sodium and potassium chloride containing Triton
MABT	maleic acid-buffered saline containing Triton
Mb	megabase
MGD	mouse genome database
MGEIR	mouse gene expression information resource
mRNA	messenger RNA
NBT	4-nitroblue tetrazolium chloride
NTE	sodium chloride–Tris–EDTA
PAC	P1 artificial chromosome
PAGE	polyacrylamide gel electrophoresis
PBS	phosphate-buffered saline
PBT	PBS containing Triton or Tween
PCR	polymerase chain reaction
PFA	paraformaldehyde
PHA	phytohaemagglutinin
PI	propidium iodide
RNase	ribonuclease
r.p.m.	revolutions per minute
snRNA	small nuclear RNA
snoRNA	small nucleolar RNA
SSC	standard saline citrate
ssDNA	single-stranded DNA
STS	sequence tagged site
T	thymine (or thymidine)
TBE	Tris–borate–EDTA
TBS	Tris-buffered saline
TdT	terminal deoxyribonucleotide transferase
TE	Tris–EDTA
TEMED	tetramethylethylenediamine
TESPA	3-aminopropyltriethoxysilane
T_m	melting temperature
T_r	reassociation temperature
TRITC	tetramethylrhodamine isothiocyanate
tRNA	transfer RNA
U	uracil (or uridine)
UTP	uridine triphosphate
UV	ultraviolet light
WWW	World Wide Web
YAC	yeast artificial chromosome

1

The theory and practice of *in situ* hybridization

DAVID G. WILKINSON

1. Introduction

The *in situ* detection of nucleic acid sequences, whether of genes on chromosomes or viruses, or of mRNA in tissues, provides a direct visualization of the spatial location of specific sequences that is crucial for elucidation of the organization and function of genes. As a consequence, methods of *in situ* hybridization have become important techniques in a number of fields, including diagnosis of chromosomal rearrangements, detection of viral infections, and analysis of gene function during embryonic development. *In situ* hybridization takes advantage of the specific annealing of complementary nucleic acid molecules, which can be DNA and/or RNA, through hydrogen bonds formed between bases attached to the sugar–phosphate backbone: adenine (A) anneals with thymine (T, in DNA) or uracil (U, in RNA), and cytosine (C) anneals with guanine (G). This base pairing underlies the formation of a double-stranded complex, in which one strand has the opposite orientation to the other with respect to the sugar–phosphate backbone (*Figure 1A*); the sequence is read from 5' to 3' (this refers to the positions of the sugar at which the phosphate residues are attached). Any nucleic acid sequence can therefore be specifically detected by use of a probe that is the 'antisense' reverse complementary sequence. An important practical asset is the speed with which such probes can be generated by the cloning of DNA sequences, amplification by the polymerase chain reaction (PCR), or the synthesis of oligonucleotides. *In situ* hybridization involves (see *Figure 1B*):

(a) Generation of a nucleic acid probe, labelled to enable subsequent detection.

(b) Preparation of chromosome spreads or fixation of tissues (which can be sectioned).

(c) Pre-treatment of tissues to increase accessibility of target nucleic acid.

(d) Hybridization of labelled probe to chromosomes or tissues.

David G. Wilkinson

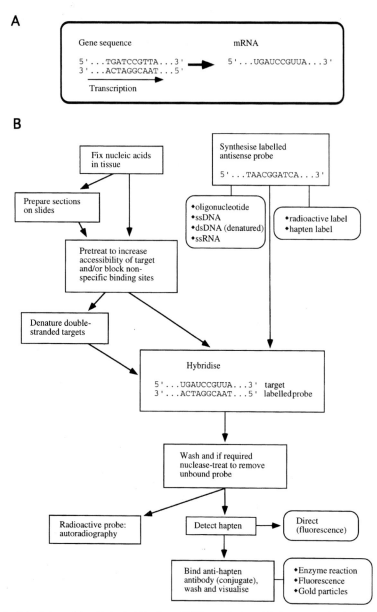

Figure 1. Principles of *in situ* hybridization. (A) The relationship between double-stranded DNA and its RNA transcript. Genomic DNA consists of two complementary strands—A paired with T, G paired with C—that are in opposite 5′ to 3′ orientations. Transcription uses the bottom strand as template, and generates RNA identical in sequence to the DNA top strand (but with U in place of T). (B) The major steps of *in situ* hybridization are illustrated (see text for more details). Note that the antisense probe is identical in sequence to the bottom strand of the genomic DNA (which is complementary to the mRNA sequence) and is written from 5′ to 3′.

2

(e) Washing under conditions that remove non-hybridized probe.

(f) Detection of the labelled probe, revealing the location of the target cellular nucleic acid.

This method can also be combined with PCR in order to amplify the target sequences and enhance detection (see ref. 1 for procedures of PCR *in situ* hybridization). A critical aspect of all the methods is that the target nucleic acid is retained *in situ*, is not degraded by nucleases, and is accessible for hybridization to the probe. Unlike the hybridization of nucleic acids in solution, the target nucleic acid is cross-linked and embedded in a complex matrix that hinders access of the probe and decreases the stability of the hybrids. Little is understood at the theoretical level of hybridization under these conditions and the procedures of *in situ* hybridization have a largely empirical basis. Similarly, it is difficult to predict non-specific binding of probe, or the efficiency of penetration of reagents used to detect probe into this complex matrix. Nevertheless, adaptations to standard hybridization techniques, with trial and error, have led to the development of reliable procedures that can be applied to many different tissues.

Many alternatives exist for the type of probe, and the methods of labelling and visualization. The major alternatives are whether to use:

• RNA probes, long DNA probes (single- or double-stranded), or oligonucleotides

• radioactive labels or non-radioactive (hapten) labels

• direct or indirect detection of the label

The method of choice depends upon what is required:

(a) Sensitivity. The sensitivity depends on the accessibility of the target, the amount and type of label included in the probe, and the method of its detection.

(b) Resolution. The resolution of the signal can vary from specific sites within a single cell, to greater than a cell diameter, and depends on the probe label and method of detection (see *Figure 2a–e*).

(c) Specificity. This is essential. The specificity of the signal depends upon the stringency of washing and the extent of similarity between the probe and sequences related to, but distinct from, the intended target.

(d) Analysis of cells or tissue sections mounted on slides, or of free-floating sections, whole tissues, or embryos (see *Figure 2a–e*).

(e) Detection of multiple genes or gene products. The physical relationship between multiple genes or chromosomal regions can be visualized by use of different fluorescent labels for detection of each sequence. For certain applications the simultaneous detection of different gene products in the same cell or tissue is important. This can be achieved by detecting several

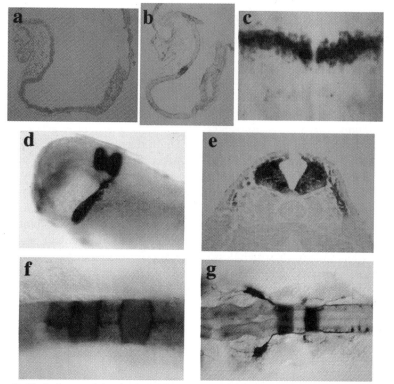

Figure 2. Techniques of *in situ* hybridization to mRNA. Examples are shown of different techniques of *in situ* hybridization to mRNA, using RNA probes for *Krox-20* gene transcripts which are expressed in one (a–c) then two (d–g) stripes in the hindbrain and in specific migrating neural crest cells (d, e). (a–c) Mouse embryos; (d, e) *Xenopus* embryos; (f, g) zebrafish embryos. (a) *In situ* hybridization with ^{35}S-labelled probe to a section. This is a double exposure photograph with the silver grains (visualized with dark-field illumination and coloured by viewing through a red filter) superimposed on the bright-field image of the toluidine blue stained section. This technique gives a resolution of greater than a cell diameter. (b) *In situ* hybridization with DIG labelled probe to a section visualized with an alkaline phosphatase (AP) colour reaction using BCIP/NBT. This technique gives a single cell resolution. (c) *In situ* hybridization with DIG labelled probe to a whole mount embryo (AP colour reaction with BCIP/NBT), photographed as a flat mount preparation of the hindbrain. This gives a single cell resolution, and reveals aspects of the spatial organization of cells (the fuzziness of the stripe) not seen in tissue sections. (d) *In situ* hybridization with DIG labelled RNA probe to a whole mount embryo (AP colour reaction with BCIP/NBT) revealing the spatial organization of expressing cells. (e) Transverse section through the whole mount hybridized embryo shown in (d) prepared after fixing the colour reaction product and embedding in paraffin wax. This reveals expression in individual cells. (f) Double *in situ* hybridization to a whole mount embryo using DIG and fluorescein labelled probes that are sequentially detected with AP-conjugated antibodies using Fast Red (*Krox-20* transcripts) and BCIP/NBT (Pax-6 transcripts) as substrates. This reveals the spatial relationship between the expression domains of these two genes. (g) *In situ* hybridization to detect *Krox-20* mRNA (AP colour reaction with BCIP/NBT) followed by immunocytochemistry to detect HNK-1 antigen (horse-radish peroxidase colour reaction with diaminobenzidine) in a whole mount embryo. HNK-1 antigen is present in specific nerves, and this technique has revealed the relationship between gene expression domains and nerve organization.

different mRNAs or by combining *in situ* hybridization with immuno-cytochemistry (see *Figure 2f* and *g*, and Chapter 5).

(f) Three-dimensional patterns. Visualization of the spatial pattern of transcripts in a tissue or embryo can provide important information. This can be achieved by computer-aided reconstruction from serial sections or by whole mount *in situ* hybridization with hapten labelled probes (see *Figure 2c* and *d*, and Chapters 4 and 5).

(g) Convenience and safety. Non-radioactive probes are more stable, safer, and give more rapid results than radioactive probes. Those not familiar with molecular biology techniques may find oligonucleotides more convenient.

This chapter discusses the theory and practice of each of the steps of *in situ* hybridization and provides a guide to the variations in technique that are described in more detail in the subsequent chapters (see also Chapters 2 and 4 for advice on general strategies for the use of oligonucleotide and RNA probes, respectively). Procedures that apply only to *in situ* hybridization at the electron microscope level or the detection of DNA sequences on chromosomes are not considered here, but are described in Chapters 6 and 7, respectively.

2. Preparation of probe

For single-stranded probes the reverse complementary sequence must be used. When probes are synthesized by transcription, this involves transcribing in the opposite direction to that used to generate mRNA from the endo-genous gene. By convention, the gene sequence is written from 5′ to 3′ as the 'top' strand that is complementary to the 'bottom' template strand, and is therefore identical in sequence to the transcribed RNA (with U residues in RNA at the position of T residues in DNA). Transcription in the opposite direction will use the top strand as template, thus synthesizing a bottom strand sequence complementary to the cellular mRNA (*Figure 1A*). When designing oligonucleotide probes (Chapter 2), it is important to remember that the antisense sequence is the *reverse* complement of the top strand, and must be written from 5′ to 3′.

In general, it is preferable to use probes in which the sequence is an exact match to the intended target, because this makes it much easier to establish that the hybridization is specific. There are situations in which this might not be possible, for example if the gene has yet to be cloned from the species being used for *in situ* hybridization. If the sequence of the gene is sufficiently similar between species, a cross-species hybridization probe can be used; for example, the nucleotide sequence corresponding to a strongly conserved poly-peptide region may be sufficiently similar, though this is difficult to predict. Cross-species hybridizations may require the stringency to be reduced since the presence of mismatched bases will reduce the stability of the hybrids (see

Section 5.1), and a problem is that under such conditions cross-hybridization to related mRNAs could occur. If the expression pattern observed is the same as seen with a homologous probe in another species, this is encouraging but still not definitive proof of specificity. In principle, this could be checked by carrying out Northern and Southern hybridizations under similar conditions, but in practice it might be simpler to clone the gene from the species under study. Cross-species hybridizations nevertheless may give useful preliminary information.

It is possible that in addition to hybridizing the intended target, even a homologous probe could bind to a closely related sequence present in another gene(s). This could occur due to the sequences having a common evolutionary origin (duplication of genes), or through serendipity. Probe design should be guided by the following points:

(a) In general, cross-hybridization is not a problem for long probes (> 100 bp) used under high stringency conditions (Section 5.1), especially after nuclease digestion to remove single-stranded probe, because it is unlikely that there would be a sufficiently long region of near-identity for the hybrid to be stable. An exception to this is when there has been little sequence divergence between related genes, because of a recent gene duplication, and/or selective pressure to maintain a particular nucleotide sequence (due to the redundancy in the genetic code, usually sufficient divergence in nucleotide sequence occurs even when the amino acid sequence is strongly conserved). Untranslated region sequences are even less likely to be conserved, and therefore are selected by some workers in preference to coding region probes. Although there have not been systematic analyses of the efficiency of such probes, some untranslated sequences might be masked by secondary structure and/or intracellular RNA binding proteins. Our experience has been that coding region probes rarely give problems, and usually much better results are obtained with a long probe that includes coding region compared to a short untranslated region probe (Section 2.2).

(b) For shorter probes (such as oligonucleotides 20–30 bases in length) the probability that another gene has an almost identical sequence is greater, though some screening for this can be carried out by searching sequence databases (Chapter 2).

(c) Poly(T) and poly(U) sequences will hybridize to the poly(A) tails of mRNAs (therefore avoid making probe by transcribing cDNAs that include poly(A) tails), and long GC-rich sequences can be 'sticky' due to the greater stability of G:C base pairs.

Two major choices need to be made for the preparation of probe: the type of nucleic acid to be used and the label to be incorporated into the probe. In addition, for certain types of probe, several alternative methods are available for probe synthesis. An important factor is the length of the probe and the means by which this is controlled depends on the type of probe and method of synthesis.

2.1 Type of probe and method of synthesis

A number of different types of nucleic acid probes can be prepared for use in *in situ* hybridization:

(a) Double-stranded DNA probes can be prepared by carrying out nick trans- lation, random priming, or PCR in a reaction in which a labelled nucleotide is incorporated. Random priming and PCR give the highest specific activities. These probes must be denatured before use. When used to detect a single-stranded target, such as mRNA, they are less sensitive than single-stranded probes, since the two strands reanneal, thus reducing the concentration of probe available for hybridization to the target.

(b) Long single-stranded DNA probes can be prepared by primer extension on single-stranded templates or by asymmetric PCR, with one of the nucleotides labelled. The former method is inconvenient, since it requires the purification of probe away from the template, and is little used for *in situ* hybridization. In contrast, PCR-based methods are much easier and probes can be synthesized from small amounts of starting material (2). Moreover, PCR allows great flexibility in the choice of probe sequences by the use of appropriate primers. Thus far, these probes have not been widely used.

(c) Oligonucleotides, typically 20–40 bases in length, can be synthesized and allow specific probes to be readily designed from published sequences (Chapter 2). These are usually labelled by incorporating modi- fied nucleotides during chemical synthesis or by adding a 'tail' of labelled nucleotides. A disadvantage of oligonucleotides is that fewer labelled nucleotides are incorporated per molecule of probe and thus they are less sensitive than longer nucleic acid probes.

(d) Single-stranded RNA probes can be synthesized by the use of an RNA polymerase (SP6, T7, or T3 RNA polymerase) to transcribe sequences downstream of the appropriate polymerase initiation site (Chapters 3 and 4). Most commonly, the probe sequence is cloned into a plasmid vector so that it is flanked by two different RNA polymerase initiation sites. This enables either sense (control) or antisense (probe) RNA to be synthesized by transcription in the same or opposite direction from the endogenous gene, respectively. The plasmid is linearized with a restric- tion enzyme so that plasmid sequences are not transcribed, since these cause high backgrounds. It is recommended to linearize with enzymes that give a 5′ overhang or blunt end, since 3′ overhangs are reported to lead to synthesis of abnormal long transcripts.

So which type of probe is best? Long single-stranded probes (RNA or DNA) are recommended for the highest sensitivity of detection of a single-stranded targets. However, oligonucleotides are sufficiently sensitive for many purposes and the sensitivity can be increased by use of a cocktail of probes against different regions of the target. Furthermore, they do not require expertise in molecular cloning techniques. For studies of closely related mRNAs, such as alternatively spliced transcripts, specific probes may be most readily obtained by the synthesis of oligonucleotides. However, if longer probes are desired, but convenient restriction sites are not available to obtain them, primers can be designed that are used to PCR amplify the desired fragments from cloned, or even genomic, DNA. These fragments can then be cloned and used to make double- or single-stranded DNA probes or RNA probes. Alternatively, they can be labelled directly without cloning; in the case of RNA probes, this is achieved by including an RNA polymerase initiation site in one of the PCR primers (3). Finally, double-stranded DNA probes are recommended for double-stranded targets, and these can be sufficiently sensitive to detect high abundance single-stranded targets.

2.2 Length of probe

Longer probes can give stronger signals because it is possible to incorporate more labelled nucleotides into them. However, probes that are too long give weaker signals because they penetrate less efficiently into the cross-linked tissue. Thus, the best strategy is to use a long sequence as template for the synthesis of probe, but to ensure that the probe molecules generated are sufficiently small to penetrate. Many procedures are based on results indicating that probes of 50–150 bases give optimal signals for hybridization to sections of certain tissues (4). However, a number of laboratories have found that RNA probes of up to 1 kb or even longer give optimal signals for *in situ* hybridization to paraformaldehyde-fixed embryos (Chapters 4 and 5). The extent to which long probes are hindered from penetrating will depend upon the nature of the tissue, the choice of fixative (see Section 3.1), and whether pre-treatments have been carried out to remove cellular protein (see Section 4). It is therefore advisable to check whether or not reduction of probe length improves signal for the tissue that you are using.

The length of probe can be controlled either during the synthesis reaction or by a subsequent partial cleavage:

(a) For nick translated DNA probes, the length is determined by the amount of DNase in the reaction.

(b) For probes labelled by random priming the length is determined by the concentration of the primer.

(c) Appropriate primers can be chosen to determine the length of probe synthesized by PCR.

(d) Long DNA probes can be partially cleaved by incubation at high temperature (2).

(e) Long RNA probes are partially cleaved by limited alkali hydrolysis (Chapters 3 and 4).

If long probes are cleaved by hydrolysis, it is important that probe size is checked since molecules that are too small will not hybridize under the standard stringency conditions, leading to low signals.

2.3 Label

Two major alternatives exist for the label to be incorporated into probe: radioactive labels which are detected by autoradiography, or hapten (non-radioactive) labels which are detected either directly, or indirectly by immunocytochemistry.

Several different isotopes have been used to radiolabel probes for *in situ* hybridization that differ with regard to the resolution of the signal, the speed of results, and the stability of the probes (see also Chapter 2). ^{3}H-labelled probes give a subcellular resolution of signal, but require long autoradiographic exposures. ^{35}S-labelled probes give a resolution of about one cell diameter, with typical autoradiographic exposure times of about one week. Whenever possible a reducing agent, such as dithiothreitol (DTT), should be included in solutions containing ^{35}S-labelled probe, to protect the sulfur from oxidation. The half-life of this isotope is 87 days, and probe should be used within a month. ^{32}P gives a poorer resolution of signal than ^{35}S and, due to the low efficiency of autoradiography, offers little advantage in the speed of results. Moreover, ^{32}P has a half-life of only 14 days and probes have to be used within a week. Finally, ^{33}P can be used, which gives a similar resolution as ^{35}S, but has a shorter half-life (25 days), and is currently more expensive. Quantitation of results is possible with radioactive probes (Chapter 2).

Hapten labelled probes have a number of assets, including safety, their high stability, rapid results, and a single cell resolution. In addition, *in situ* hybridization can be carried out in whole mount (Chapters 4 and 5). The most sensitive methods involve incorporating a modified nucleotide that has a hapten group for which a specific antibody or other binding protein is available. Following hybridization and washing, the tissue is incubated with hapten binding protein coupled to an enzyme, and then the signal is produced by using a substrate for this enzyme that yields an insoluble, coloured product. The range of haptens that can be used is currently limited by the commercial availability of antibodies or binding protein. The most commonly used haptens are digoxigenin or fluorescein, for which conjugated antibodies are available, or biotin which can be detected with streptavidin conjugates.

Fluorescent labelling, either of the probe or of the anti-hapten antibody, offers the advantage of using different fluorescent tags to simultaneously detect different sequences, and is widely used for chromosomal *in situ* hybridization (Chapter 7). However, this is currently little used for detecting mRNA because of the lower sensitivity. A further alternative is the use of gold-coupled antibodies, as used for *in situ* hybridization at the electron microscope level (Chapter 6).

The choice of probe label depends upon the requirements for sensitivity and resolution of signal. ^{35}S-labelled probes are, at present, more sensitive than hapten labelled probes, and may therefore be the choice for detecting low abundance transcripts (Chapters 2 and 3). If single cell resolution and/or three-dimensional visualization of gene expression is required, or safety and convenience is an important consideration, then the use of hapten labelled probes is the method of choice (Chapters 4 and 5).

3. Preparation of tissue

The procedure used to prepare tissues is determined by the nature of the tissue sample and the specific requirements of the experiment. Cells in tissue culture can be grown directly on slides or coverslips, to which they are then fixed, or cell suspensions can be spun down onto subbed slides prior to fixation. Tissues can be sectioned or hybridized in whole mount. The major choices that need to be made are the fixative, the methods of tissue embedding, and the subbing of slides for the attachment of cells or tissue sections.

3.1 Fixation

Cross-linking fixatives, such as aldehydes, give a greater accessibility and retention of cellular RNA than precipitating fixatives. The conditions for fixation are a compromise: stronger fixation yields a better preservation of cellular morphology, but the increased cross-linking lowers the accessibility of the target. 4% paraformaldehyde, 4% formaldehyde, or 1% glutaraldehyde are most commonly used, the latter giving more extensive cross-linking. Fixation is usually carried out at 0–4°C to inhibit endogenous ribonucleases. The length of fixation depends on the size of the sample: for isolated cells, 20 min is sufficient; for tissues of less than 1 mm thickness, 2–3 h are sufficient; for tissues of less than 0.5 cm thickness, overnight is sufficient; for larger tissues, perfusion may be necessary. Alternative fixatives and protocols may be required for combining *in situ* hybridization with immunocytochemistry.

3.2 Embedding and sectioning of tissue

Tissue sections can be prepared either on a cryostat or, after embedding in a matrix, on a microtome. Excellent results can be obtained with cryostat sectioning, and this method allows the possibility of fixing after sectioning, so that different fixatives for immunocytochemistry and hybridization can be used on adjacent sections. However, the sectioning of embedded tissue has several advantages, including the better preservation of tissue morphology, the ability to cut thin sections, and the ease with which the sample can be accurately orientated and serial sections obtained. Paraffin wax is the most popular embedding medium since sections of down to 1 μm can be cut and it can be fully dissolved and washed out from the tissue prior to hybridization. Thin sections have the potential disadvantage that the signal will be lower, and 6–10 μm sections are commonly used. It is also possible to use thicker sections which can be mounted on slides or hybridized as free-floating sections (Chapter 2). Embedding in plastic allows thinner sections to be cut, as required for electron microscopy (Chapter 6), but masks much of the target nucleic acid.

3.3 Subbing of slides

Slides are subbed so that cells or tissue sections adhere and are retained during the subsequent steps of *in situ* hybridization. Many different methods are suitable, including the use of chrome-alum gelatin, poly-lysine, or amino-propyltriethoxysilane (TESPA). Many workers find that TESPA gives the strongest adherence. Alternatively, slides coated with positively charged groups can be purchased that give strong adherence of sections (for example, Superfrost Plus slides from BDH).

4. Pre-treatments of tissue

Prior to hybridization, the tissue is subjected to a series of pre-treatments that increase the efficiency of hybridization and/or decrease non-specific background:

(a) Whole tissues or sections are usually treated with organic solvents (ethanol or methanol) to permeabilize the cells by removing lipid membranes. For wax-embedded tissues, this is in any case required for tissue embedding and the removal of wax before hybridization.

(b) In order to increase the accessibility of the target RNA, the tissue is treated with protease (usually proteinase K) to partially digest cellular proteins. This is usually followed by a refixation step, since otherwise disintegration of the tissue will occur. Protease treatment substantially improves the signal obtained with long (> 100 bp) probes, but is not required when oligonucleotide probes are being used. The optimal extent of protease treatment should be ascertained, since it differs between tissues. A problem for the processing of whole embryos is that since it will take some time for the protease to penetrate, internal tissues will be less efficiently digested than the surface. This will further exaggerate the more efficient detection of signal in surface layers due to limited penetration of probe and detection reagents. An alternative is to treat the tissue with a mixture of detergents (5); this is easier to control but is less effective than

11

protease treatment. In some protocols, treatment of sections with heat or acid is carried out in order to remove protein.

(c) Non-specific binding of probe to positively charged amino groups can be prevented by acetylation of these residues with acetic anhydride. However, in some protocols this step has no effect on background and is optional.

(d) For whole mount hybridizations, tissues are pre-hybridized by incubation in hybridization solution lacking probe in order to block non-specific binding sites. For tissue sections, dehydration is carried out so that the probe is placed onto and absorbed by the dried section. Most workers have found that a pre-hybridization step is not necessary for sections, and has the problem that the pre-hybridization solution substantially dilutes the small volume of probe, leading to decreased signal.

(e) For double-stranded DNA targets, a heat denaturation step is carried out.

It is essential to avoid degradation of the target nucleic acid by nucleases during these pre-hybridization steps. This is especially important if RNA is the target, since this is highly susceptible to degradation by ribonucleases. Ribonucleases should be inactivated by DEPC treating and/or autoclaving solutions, and precautions taken that containers used for the processing of tissues are not contaminated by ribonucleases (autoclave or bake these as necessary, and wear gloves for handling them).

5. Hybridization and washing

Following the pre-treatments, hybridization is carried out under optimal conditions for the annealing of probe to the target nucleic acid in the tissue. The tissue is then washed to remove non-specifically bound probe.

5.1 Optimal conditions for hybridization

The melting temperature, T_m, at which 50% of the hybrids are dissociated is influenced by a number of factors:

(a) Nature of the probe and target. RNA:RNA hybrids are more stable than RNA:DNA hybrids, which are more stable than DNA:DNA hybrids.

(b) Length of the probe. Longer probes form more stable hybrids.

(c) Base composition of the probe. There are two hydrogen bonds in each A:T/U base pair, and three hydrogen bonds in each G:C base pair. Consequently, the greater the GC content, the higher the melting temperature.

(d) Extent of sequence identity between the probe and target. If the probe is divergent in sequence from the target, it will form less stable hybrids than homologous probes. The effect on the T_m is greater for short probes.

(e) Composition of the hybridization/washing solution. Higher concentrations of monovalent cations, such as sodium ions, increase the stability of hybrids. Formamide decreases the T_m, enabling lower temperatures to be used to achieve a specific stringency, and is usually included during the prolonged period of hybridization to avoid any tissue disintegration that might occur at a higher temperature.

Hybridization experiments have lead to a number of formulae that approximately predict the T_m. All of these apply to homologous nucleic acids hybridizing in solution, whereas hybrids formed with the cross-linked nucleic acids present in fixed tissue are less stable, presumably because steric hindrance prevents the annealing of probe along its full-length. RNA:RNA hybrids formed during *in situ* hybridization have a melting temperature about 5°C lower than hybrids formed in solution (4). In the formulae given below, fraction G + C is the amount of G + C residues expressed as a fraction of the total residues in the sequence.

(a) DNA:DNA hybrids (DNA probes longer than 22 bases):

$T_m = 81.5 + 16.6$ log (molarity of monovalent cations) + 41 (fraction G + C) – 500/(length of probe in bases) – 0.62 (%formamide).

(b) RNA:RNA hybrids (for RNA:DNA hybrids, the last term is replaced by $0.5 \times$ %formamide):

$T_m = 79.8 + 18.5$ log (molarity of monovalent cations) + 58.4 (fraction G + C) + 11.8 (fraction G + C)^2 – 820/(length of probe in bases) – 0.35 (%formamide).

(c) Hybridization of DNA 11–22 bases in length to DNA in 0.9 M salt:

$T_m = 4$ (number of G + C residues) + 2 (number of A + T residues).

The kinetics of *in situ* hybridization are less well understood, especially since they are influenced by the accessibility of the target which will depend on the extent of cross-linking of the tissue, the size of the probe, and, for mRNA, possibly on secondary structure. The following factors are important:

(a) Concentration of probe. For RNA:RNA hybrids, increasing concentrations of probe, well above theoretical probe excess, lead to greater signals after completion of the hybridization reaction. Saturation does occur, however, and an excessive probe concentration will lead to increased background. The effective probe concentration and rate of hybridization can be increased by using a volume exclusion agent such as dextran sulfate.

(b) Stringency of hybridization. For RNA:RNA hybrids, hybridization to sections is almost complete after 5–6 h at a temperature 25°C below the T_m (4). Based on this, hybridization is generally carried out under these moderate conditions and, if required, discrimination between related sequences achieved by high stringency washing. However, high stringency hybridization is used in many protocols, and presumably the longer period generally used for hybridization (usually overnight) compensates for the lower rate of hybridization under these conditions.

5.2 Post-hybridization washing

Washing of the hybridized sections is carried out to remove probe that has bound to any sequences related to, but distinct from, the intended target or non-specifically to other cellular components. The critical requirement is to use appropriate stringency conditions in order to remove only the non-specifically bound probe, and typically washing is carried out at moderate or high stringency (up to several °C below the T_m). It is important to be aware that short or AT-rich probes will give poor signals if washed at too high a stringency. For RNA probes, a ribonuclease step can be carried out in order to degrade single-stranded probe, leaving only the double-stranded hybrid intact. In addition to removing non-hybridized probe, this will degrade single-stranded regions of any probe molecules that have formed a heterologous hybrid with a similar sequence. It can therefore be used to increase the specificity of the signal for homologous target, though in practice this can usually be achieved by high stringency washing. Care should be taken that agents that inhibit the ribonuclease are not present during this step; for example, reducing agents (used to protect ^{35}S-labelled probes from oxidation) and formamide must be removed. However, ribonuclease treatment decreases signal significantly, probably because due to steric hindrance the probe has not hybridized along its full-length even to homologous target, and the single-stranded regions are digested. This treatment therefore is only carried out if it is essential.

6. Visualization of signal

The final step of *in situ* hybridization is the detection of the labelled probe in the tissue. The method for this depends upon the type of label that has been incorporated into probe.

6.1 Radioactive probes

For ^{35}S-labelled probes, a low-resolution signal can be obtained by placing the slide adjacent to X-ray film and overnight exposure. This can be useful for obtaining rapid results when optimizing conditions or troubleshooting problems, but the resolution is too low to be useful for most purposes. A much greater resolution is obtained by dipping the slide in liquid nuclear track emulsion, which is then dried, exposed at 4 °C, and developed. Several factors affect the quality of the results obtained by this method (discussed in detail in ref. 6):

(a) The grade of emulsion. Coarser grades of emulsions are more sensitive, but give a poorer resolution of signal.

(b) The thickness of the emulsion. For ^{35}S, a thicker layer of emulsion gives a greater signal, but poorer resolution.

(c) Storage and handling of emulsion. Emulsion must be kept away from penetrating radiation to avoid a high background of silver grains. In addition, silver grains are generated by mechanical stress, so the liquid emulsion should be treated gently, and the drying onto slides should not be too rapid.

(d) Exposure conditions. The temperature and humidity conditions affect the efficiency of autoradiography.

(e) Exposure time. Overexposure should be avoided since this leads to a reduction in the signal-to-noise ratio.

(f) Conditions for developing. The use of a higher temperature, more agitation of the slides, or a longer time for developing yields larger silver grains, but a lower resolution.

After developing, the tissue is counterstained, usually with a nuclear stain such as toluidine blue, and mounted under a coverslip. High densities of silver grains can be observed under bright-field illumination on a light microscope, but a more sensitive means of visualizing silver grains is to use dark-field illumination. A problem with this is that the tissue staining is not clearly seen under dark-field illumination, so it can be difficult to precisely correlate the signal with specific cells. However, it is easy to superimpose the signal and tissue by taking a double-exposure photograph of the silver grains under dark-field, coloured by the use of a red filter, upon a bright-field image of the stained tissue (*Figure 2a*).

6.2 Hapten labelled probes

The location of hapten labelled probes can be visualized by fluoresence microscopy, directly if the probe is fluorescently labelled, or indirectly with a conjugated antibody (Chapter 5). These techniques work beautifully for chromosomal *in situ* hybridizations and give great flexibility for the simultaneous detection of multiple sequences (Chapter 7), but currently are only sensitive enough for the detection of relatively abundant mRNAs. A more sensitive method is to amplify the signal by using an enzyme-conjugated antibody or binding protein, for which chromogenic substrates are available that yield an insoluble coloured product. The most commonly used reagents are anti-DIG or anti-fluorescein antibody conjugated with alkaline phosphatase (AP), using NBT/BCIP as substrate, as the stability of this enzyme allows great amplification of the signal. A variety of other substrates are available that yield red, magenta, or fluorescent green products (see Chapter 5), but these are less sensitive than NBT/BCIP. β-Galactosidase is also a stable

15

enzyme and with X-Gal as substrate is commonly used in other contexts, but although a streptavidin conjugate has been used successfully (7), this has not been widely used. Horse-radish peroxidase is less stable, and can only be used for short chromogenic reactions.

Prior to incubation with diluted anti-hapten enzyme-conjugated antibody, the hybridized tissues are pre-incubated with a solution containing detergent and other reagents that block non-specific binding of the antibody. In some protocols, the antibody is pre-absorbed with a tissue powder prepared from the system under analysis in order to remove any non-specific binding activity; however, this step is often not required. The tissue is then incubated with the antibody at an optimal concentration; if the antibody is too dilute the signal will be weak, but excessive concentrations will cause a high background. Extensive washing with buffer containing detergent, such as 0.1% Triton X-100, is then carried out to remove unbound antibody. Higher concentrations of detergent (e.g. 1% Triton X-100) can give lower non-specific binding of the antibody, but may considerably decrease the signal strength. Finally, the colour reaction is carried out by equilibrating the tissue with the chromogenic substrate that is converted by the enzyme into an insoluble coloured product. It is important that there are not any endogenous enzymes that can also generate this product. In the case of alkaline phosphatase, levamisole can be included in the reaction as this strongly inhibits many phosphatases, but not the intestinal phosphatase that is conjugated to the antibody. For many tissues this is not necessary, whereas in other cases there may be endogenous phosphatases that are not affected by this inhibitor; these can usually be inactivated by heat or acid treatment prior to incubation with anti-hapten antibody (as used for sequential detection of mRNAs with AP-conjugated antibody— see Chapter 5). If peroxidase is used for the colour reaction, endogenous peroxidases may need to be inactivated by pre-incubation with a hydrogen peroxide solution. However, the high temperature used during hybridization and washing should inactivate endogenous peroxidases, and many phosphatases. Some protocols include a hydrogen peroxide incubation during the pre-treatment even though AP-conjugated antibody is to be used; possibly this treatment also inactivates phosphatases.

6.3 Double staining techniques

The detection of multiple gene sequences on chromosomes is invaluable for the analysis of their physical organization, and this can be achieved by the use of anti-hapten antibodies linked to different fluorochromes (Chapter 7). The detection of several gene products is very useful for relating the expression of a particular gene to specific cell types or spatial domains marked by the expression of a previously characterized gene. Ascertaining whether two genes have overlapping or adjacent expression domains can also give valuable clues regarding possible interactions between their products. It is possible to combine *in situ* hybridization using radioactive probes with other detection methods, but a more convenient multiple detection can be achieved by labelling probes with different haptens that are then detected with enzyme-conjugated reagents (Chapter 5). Two different enzymes can be used for the chromogenic reactions, but due to the high stability of alkaline phosphatase, the most commonly used method is to detect the probes sequentially with AP-conjugated antibodies, using substrates that yield different coloured products. Alternatively, *in situ* hybridization can be combined with immunocytochemistry to detect specific proteins or other antigens (Chapter 5). For this to work, the detection of one gene product must not compromise the subsequent detection of the other. If the antigen survives the *in situ* hybridization procedure (some protein antigens will even survive the proteinase treatment), then it is simplest to detect the RNA then the antigen. If the steps required for *in situ* hybridization are incompatible with detection of antigen, it is possible to reverse this order, but this will require strict precautions to minimize RNase activity during immunocytochemistry. It is also possible to use other fixatives in this procedure, as may be required to preserve antigenicity (8). A further method for the simultaneous detection of gene products is to use transgenic animals that express a reporter gene, most commonly lacZ, in specific tissues of the embryo; fixed embryos are first stained for the reporter gene and then *in situ* hybridization carried out.

A limitation of these methods is that with the current range of available substrates it is only possible to detect overlapping gene expression under optimal conditions. The main problems are that one colour can easily mask another, and/or that the precipitated product prevents other reagents from penetrating. In addition, most of the available substrates are less sensitive or give higher non-specific staining than NBT/BCIP, so can only be used under optimal conditions. One way around this is to use substrates that yield in-soluble fluorescent products, but although several are available (Chapter 5) they are currently not ideal. A further possibility is to use fluorochrome-conjugated antibodies if they are sufficiently sensitive. It is to be hoped that these limitations will be solved by the development of new enzyme substrates and/or more intense fluorochromes and sensitive imaging techniques. In the future, it may even be possible to generate modified nucleic acids that will enable the spectroscopic detection of hybrids *in situ* in living tissues.

In situ hybridization can also be combined with techniques that reveal aspects of cell physiology or organization, such as the detection of a cell lineage tracer (9, 10), of cell proliferation by the *in vivo* incorporation of BrdU into DNA (11), or of apoptosis by the end-labelling of fragmented DNA (12). The design of such protocols depends upon whether one detection is affected by the other. For example, if tracing cell lineage using the fluorescent molecule DiI that intercalates into lipid membranes, photoconversion of this signal into a permanent stain has to be carried out before *in situ* hybridization.

7. Controls and troubleshooting

Signal can arise not only at the sites of hybridization of probe to target but also elsewhere. Thus, it is important that appropriate controls are carried out to test the specificity of the signal. There are several possible causes of non-specific signals.

1. Cross-hybridization to other nucleic acid sequences.

    (a) Precautions should be taken in the design of probe to minimize the possibility of cross-hybridization to related sequences (see Section 2). Test whether the same pattern is observed using probes for distinct regions of the transcript, one of which does not include any conserved domains of the coding region.

    (b) If moderate stringency hybridization is being used, the possibility that cross-hybridization has occurred under these conditions can be tested by observing whether the same expression pattern is detected at high stringency, or if using an RNA probe, after a post-hybridization RNase treatment. However, the signal will be lower, so it can sometimes be difficult to ascertain whether or not weak signals seen under less stringent conditions are due to homologous hybridization.

    (c) Ascertain that the probe does not cross-hybridize under equivalent stringency conditions to other genes or mRNAs on genomic Southern blots or Northern blots. The latter is usually of limited value as it requires prior knowledge of whether transcripts from different genes can be distinguished by size on Northern blots.

2. Non-specific binding of probe or detection reagents to tissues (or for sections, to the slide).

    (a) A frequently used control for specific hybridization to mRNA is to use a sense strand probe which should not specifically anneal to the target RNA (although some genes are transcribed on both strands, in which case this 'control' will detect a specific RNA). An equally valid alternative is to use a probe from a gene expressed in a distinct pattern, or, for oligonucleotides, to use a scrambled sequence as probe.

    (b) A pre-hybridization treatment of the tissue with RNase will remove the RNA target. If signal is still generated, then this is likely to be due to binding of probe to other cellular components. This control is suitable for DNA probes (Chapter 2), but if RNA probes are used care has to be taken that residual RNase does not degrade the probe.

    (c) Inclusion of a tenfold excess of unlabelled probe in the hybridization reaction will drastically decrease the specific signal, but will have little effect on any non-specific binding to other cellular components.

    (d) If a signal is generated after omitting the probe, this may be due to non-specific binding of the detection reagent (but see below).

3. Non-specific generation of signal.

   (a) For radioactive probes, signal can be generated due to mechanical stress or chemography of the photographic emulsion. Such signal will still be present after omitting probe or using a control probe.

   (b) For hapten labelled probes detected with enzyme-conjugated antibody, signal could be generated by endogenous enzymes. This can tested by controls in which the probe and antibody are left out.

Below are listed some of the most common artefacts and possible cures:

(a) Background over the entire slide (radioactive probes). This is most frequently due to mistreatment of the emulsion. Check the emulsion by dipping a blank slide. A high background can appear at the position of the coverslip edge due to the drying-up of probe during the hybridization— avoid having tissue sections near the edge.

(b) Background around edge of tissues (radioactive probes). Silver grains can be produced around the edge of tissues by mechanical stress during drying. Dry the emulsion more slowly by placing on a cooled plate.

(c) Uniform background over the tissue. This is most frequently caused by non-specific binding of the probe, which could be due to:

   (i) For RNA probes, that plasmid sequences have been transcribed.

   (ii) The probe is too large or too small.

   (iii) The probe concentration is too high (or so low that the background becomes a limiting factor after a long exposure/colour reaction).

   (iv) The stringency of hybridization and washing is too low.

   (v) If post-hybridization digestion with ribonuclease is being used for RNA probes, this treatment is not working.

   (vi) The probe is binding to a widely distributed cellular component. It may be worth testing a different region of the same gene. Avoid using probes that are highly GC-rich or include poly(T) or poly(U) (i.e. the reverse complement of the poly(A) tail of cDNA clones).

(d) Background over specific tissues. Certain tissues seem to be more 'sticky' than others. Trying a different probe, or adjusting the probe concentration, hybridization, and/or washing conditions may help.

(e) Background with hapten labelled probes could be due to:

   (i) Too high a concentration of anti-hapten antibody.

   (ii) Insufficient blocking of non-specific binding sites before hybridization and/or detection with antibody. Use stronger conditions for blocking or washing (such as higher detergent concentration).

   (iii) The presence of endogenous enzymes that can catalyse the signal detection reaction. The use of specific inhibitors (e.g. levamisol for

alkaline phosphatase) and/or incubation of the tissue at high temperature or with acid can be used to inactivate certain endogenous
enzymes.

(iv) With biotin labelled probes, the presence of endogenous hapten in
the tissue can cause a high background. Switch to a different hapten.

(v) For whole mount hybridizations, trapping of probe or antibody in
enclosed cavities can cause background—open these up prior to
hybridization using forceps or needles.

(f) Low signal could be due to:

(i) Degradation of the target. Nuclease contamination must be avoided
in the solutions and containers used for the pre-hybridization steps,
and tissues must be fixed quickly.

(ii) Tissue fixation and/or protease treatment is not optimal.

(iii) Labelling of the probe has not been efficient. Check that there is no
problem with the labelled nucleotide (try a fresh batch) and that it is
present at the correct concentration in the labelling reaction.

(iv) The probe (after size reduction) is too short or too long.

(v) The probe concentration is too low.

(vi) The stringency of hybridization and/or washing is too high.

(vii) A low abundance of the target. The limiting factors will be sensitivity
of the method and the level of non-specific background (see above).
If possible, try using as long a probe (prior to size reduction) as is
available.

8. Photography

The photography of *in situ* hybridization data is a crucial step, since there is
little point in obtaining beautiful data if they are not recorded well. Low
power photographs can be taken using a dissection microscope, but a compound microscope is required for high power pictures. It is essential to have a
microscope equipped with the appropriate lenses and filters: for example, differential interference contrast (DIC, Nomarski) optics if you wish to visualize
cells without counterstaining, epifluorescent illumination for visualizing fluorescent conjugates or enzyme products, or dark-field illumination for visualizing silver grains. Tissue sections are mounted under a coverslip using an
organic solvent-based mounting agent, or if these dissolve the signal, using
70% glycerol. Whole tissues can be mounted in 70% glycerol, and sometimes
it can be useful to partially dissect the tissue such that the signal is visualized
easily (see *Figure 2c* and Chapter 4). The correct film should be used: for
bright-field and dark-field images, high-resolution films with an ASA of 64 or
100 are widely used; for low intensity fluorescent images it is preferable to use

faster films. Black and white photographs are often taken using negative film. For colour photography it is generally better to use slide film, since these can be used directly for presentations and it is easier to obtain the correct colour balance if using commercial printers. It is very important to learn how to set-up the microscope correctly, and how to take photographs with the correct exposure in which the subject is appropriately framed and oriented, and in focus. Excellent advice can be found in the article by Stern (13). The images can also be digitally recorded, either directly from the microscope or by scanning a photographic slide, and this greatly enhances the generation of figures for publication.

Acknowledgements

I am very grateful to Angela Nieto, Romita Das Gupta, Leila Bradley, and Qiling Xu for the data shown in *Figure 2*.

References

1. Herrington, C. S. and O'Leary, J. (eds) (1997). *PCR in situ hybridization: a practical approach*. IRL Press, Oxford.
2. Tautz, D., Hulskamp, M., and Sommer, R. J. (1992). In *In situ hybridization: a practical approach* (ed. D. G. Wilkinson), 1st edn, p. 61. IRL Press, Oxford.
3. Frohman, M. and Martin, G. (1989). *Technique*, **1**, 165.
4. Cox, K. H., DeLeon, D. V., Angerer, L. M., and Angerer, R. C. (1984). *Dev. Biol.*, **101**, 485.
5. Rosen, B. and Beddington, R. S. P. (1993). *Trends Genet.*, **9**, 162.
6. Rogers, A. W. (1979). *Techniques of autoradiography*. Elsevier, Amsterdam.
7. Conlon, R. A. and Herrmann, B. G. (1993). In *Methods in enzymology* (ed. P. M. Wassarman and M. L. DePamphilis), Vol. 225, p. 373. Academic Press, San Diego.
8. Brahic, M. and Ozden, S. (1992). In *In situ hybridization: a practical approach* (ed. D. G. Wilkinson), 1st edn, p. 85. IRL Press, Oxford.
9. Izpisua-Belmonte, J. C., De Robertis, E. M., Storey, K. G., and Stern, C. D. (1993). *Cell*, **74**, 645.
10. Nieto, M. A., Sechrist, J., Wilkinson, D. G., and Bronner-Fraser, M. (1995). *EMBO J.*, **14**, 1697.
11. Myatt, A., Henrique, D., Ish-Horowicz, D., and Lewis, J. (1996). *Dev. Biol.*, **174**, 233.
12. White, K., Grether, M. E., Abrams, J. M., Young, L., Farrell, K., and Steller, H. (1994). *Science*, **264**, 677.
13. Stern, C. D. (1993). In *Essential developmental biology: a practical approach* (ed. C. D. Stern and P. W. H. Holland), p. 67. IRL Press, Oxford.

Oligonucleotide probes for *in situ* hybridization

MARCUS RATTRAY and GREGORY J. MICHAEL

1. Introduction

Oligonucleotide probes are convenient for determining the sites of gene expression using *in situ* hybridization. Since the probes are synthetic, their use does not require much specialized equipment for molecular biology, the growth of bacteria, or the purification of nucleic acids. Oligonucleotide probes can be designed to form complementary base pairs with a target sequence, that can be RNA or DNA. In this chapter we consider binding of oligonucleotide probes to RNA targets only. The probe is labelled so that its presence on tissue sections can be detected. Probe is applied to the tissue under conditions that hybridization takes place, and then the sections are washed to remove non-specific binding. After these steps have taken place, the presence of the probe can be detected by autoradiography (for radiolabelled probes) or by histochemistry (for non-radiolabelled probes). It is possible to use oligonucleotide probes to measure the level of gene expression in individual cells or in tissues using appropriate image analysis techniques.

This chapter addresses the major issues in the use of oligonucleotide probes: their design, purification, and labelling as well as protocols for their use and detection. We also outline some basic techniques, which may be useful for those with little training in molecular biology. The protocols have worked well in our laboratories for a number of years, almost all on neuronal tissue; they should serve as a guide only, since all techniques for successful *in situ* hybridization are empirical.

2. Design of oligonucleotide probes

2.1 Theoretical considerations

2.1.1 General considerations

Important considerations for an oligonucleotide probe are its *specificity and sensitivity*:

(a) The specificity is determined by the ability of the oligonucleotide to bind to the sequence of interest, and no other sequence. Since only a fraction of all expressed genes are currently known it is not yet possible to be certain that an oligonucleotide is entirely specific. It is possible, by using genome database searches, to maximize the chances of success.

(b) The sensitivity of an oligonucleotide probe will depend on its ability to bind to the sequence of interest under the experimental conditions used, and also on how and to what extent it is labelled. Oligonucleotide probes may be able to bind to themselves rather than the target molecules. It is therefore important to ensure that the oligonucleotide lacks self-complementary regions, since if the oligonucleotides bind to each other this will reduce the amount of oligonucleotide available to bind to the target mRNA, and therefore the strength of the signal.

The binding of an oligonucleotide probe to its target is influenced by several factors, including the length and GC content (see Chapter 1). These factors determine the melting temperature (T_m) at which 50% of the hybrids dissociate. A number of equations purport to approximate the binding of oligonucleotides to their targets (1):

$T_m = 81.5 + 16.6 \log[Na^+] + 41$ (mole fraction G + C) – (500/length of oligonucleotide)

$T_m = 79.8 + 18.5 \log[Na^+] + 58.4$ (mole fraction G + C) + 11.8 ((mole fraction G + C)^2) – (820/length of oligonucleotide)

$T_m = 4$ (G + C content) + 2 (A + T content)

These equations can be used to choose a length and GC content of a probe that will maximize specific binding under a given set of experimental conditions. However, we find in practice that these equations do not truly represent hybridization of oligonucleotide probes to mRNA in tissue sections (2). We therefore present some rule-of-thumb guide-lines that work well for the *in situ* hybridization protocols outlined below.

2.1.2 Choosing length and GC content

Oligonucleotides with a higher GC content will produce stronger hybrids than oligonucleotides of an equivalent size with a lower GC content. However, empirical considerations dictate that the GC content should not be greater than 60%, since probes with a high GC content tend to produce high non-specific binding. When designing oligonucleotides, we select oligonucleotides with a GC content of 45–55% (3).

Longer probes produce more stable hybrids and allow higher temperatures and lower salt concentrations to be used during the *in situ* hybridization experiments. Factors which limit the total length of the oligonucleotide include the ability of the oligonucleotide to penetrate tissue sections, the increased chance of there being self-complementary regions in a longer oligonucleotide, and cost. Short oligonucleotides will not bind well using convenient temperatures and need to be washed at low temperatures: under these conditions, non-specific hybridization can become a major problem. We have successfully used oligonucleotides ranging in size from 20–55 bases, but opt for oligonucleotides in the size range 32–36 bases. The protocols described below have been optimized for oligonucleotides of this length.

2.1.3 Self-complementarity

Self-complementary sites are common in nucleic acid sequences. Even short oligonucleotides are capable of binding to themselves (hairpin loop formation) and binding to other oligonucleotides (stacking sites). Self-complementary sites will reduce the effective probe concentration applied to tissue sections, since there will be competition between the probe binding to the target and binding to itself. A probe with self-complementary sites is therefore more likely to be less sensitive than a probe lacking such sites.

Self-complementary sites can include mismatches (see *Figure 1* for examples). Some self-complementary sites are easy to see by eye, for example AAAA may bind to TTTT in a different part of a sequence (this is an example of a four base stacking site with 100% complementarity). However, others may be more difficult to spot, for example *Figure 1H* shows an example of a 11 base self-complementary site with 64% complementarity, that contains within it two separate five base sites each with 80% complementarity.

It is practically impossible to design an oligonucleotide without any stacking sites at all, but it is important to minimize them. In our laboratory we prefer to use oligonucleotides with sequences that contain no more than two stacking sites that are three bases in length and have 100% complementarity; no stacking sites of 100% complementarity over three bases in length; and no more than two stacking sites which are five bases in length with > 80% complementarity.

2.1.4 Specificity and mismatches

It is important that the probe binds to the target of interest, and no other. Since only a fraction of all gene sequences are presently in the EMBL and Genbank databases, it is not possible to know that a particular sequence is unique to the chosen target. Despite this, database searches often reveal unacceptable levels of similarities with a number of gene sequences. For certain classes of genes, for example those which encode members of superfamilies of proteins (G protein coupled receptors, protein kinase C, etc.), or proteins which possess functional domains that are conserved between proteins, it is best to choose regions of the gene sequence that are not part of the coding regions for these conserved domains.

Computer database searches will allow rapid assessment of the similarity to sequences in the database. Generally, we have found that for an oligonucleotide of around 32–36 bases, if there are four or more mismatches with

Oligonucleotide sequence: 5' CGATTGAAGCGATCTTAGCTAAGTCCCTACATCG 3'

5' and 3' positions	stack length, % complementarity
A	
1 CGATT |||| G 11 GCGAA	5, 80
B	
1 CGATTGAAGCGAT CTTA |||| ||||| | G 34 GCTACATCCCTGA ATC	13, 54 (4, 100)
C	
2 GATTG A |||| A 14 CTAGC G	5, 80
D	
2 GATTGAAGC |||||| 19 CGATTCTAG	9, 56
E	
2 GATTGA AGCGAT |||| C 25 CTGAAT CGATT	6, 67

5' and 3' positions	stack length, % complementarity
F	
3 ATTGA AGCGATC ||| T 28 TCCCT GAATCGAT	5, 60
G	
4 GAAG ||| 33 CTAC	4, 75
H	
5 TGAAGCGATCT TA |||||| | G 31 ACATCCCTGAA TC	11, 64 (5, 80) x2
I	
5 TGAAG C |||| G 17 ATTCT A	5, 80 (3, 100)
J	
7 AAG C ||| G 16 TTC TA	3, 100
K	
8 AGCGA T ||| C 20 TCGAT T	5, 80 (3, 100)

Figure 1. Self-complementary regions in oligonucleotides. Examples of self-complementary sites involving the 5' end of an oligonucleotide are shown derived from the *stemloop* program in *GCG* (4). Oligonucleotides are drawn as if they have formed stemloop structures, i.e. are folded back on themselves. However, the same sites will participate in 'stacking'—the binding of one oligonucleotide to another. The numbers on the left indicate the bases (top 5' position, bottom 3' position) in the oligonucleotide that participate in stacking. Vertical lines indicate base pairing. The numbers to the right indicate the number of bases and the % complementarity in the self-complementary region. We would not synthesize an oligonucleotide which had greater than two sites of three bases with 100% complementarity (3, 100), with greater than two sites of five bases with 80% complementarity (5, 80), or one or more sites of four bases with 100% complementarity (4, 100).

non-target genes, and these mismatches are scattered more or less evenly throughout the sequence, then non-specific binding should not be a problem using the protocols described below.

2.2 Design in practice

2.2.1 Choosing gene-specific stretches

The availability of powerful computers has meant that the process of oligo-nucleotide probe design should be relatively simple. However, due to the diversity of computer programs available, problems often arise because DNA sequences are stored in different formats. It is simple to use the Internet to identify probes which are specific to the gene of interest (*Protocol 1*). However, it is more difficult to choose oligonucleotides that lack self-complementary regions.

Protocol 1.   Designing oligonucleotides

A. *Internet-based method*

1. Find the sequence of interest. Using the Internet, nucleotide sequences can be found through the National Center for Biotechnology Informa-tion server (http://www.ncbi.nlm.nih.gov/). Sequences are obtained using the *Entrez* program (http://www.ncbi.nlm.nih.gov/Entrez/) by use of a key word search. The protocol for down-loading a sequence is simple for those familiar with a Web browser. DNA sequences can be saved as simple text files and can be edited using a word processor program or text editor. When using a word processor program, choose a non-proportional spacing font (e.g. courier, courier new, or letter gothic) so that the formatting of the sequence is preserved.

2. Check the details of the sequence. Each Genbank sequence has a feature table. Look in this table to check that the sequence you have down-loaded is a cDNA (as opposed to a genomic DNA sequence) without introns. Identify the coding region (CDS).

3. Choose regions of interest. A 'BLAST' search (5) can be performed to compare the sequence of interest with all known gene sequences in the database, or a subset of those sequences (http://www.ncbi.nlm.nih.gov/BLAST). From the word processor program, cut about four separate regions of around 60 bases from the CDS[a] and edit out the spaces. Save each of these sequences into a new word processor file. Paste these sequences into the program and start the BLAST search. This will tell you whether there are strong regions of homology with other genes, and some sequences might be discarded at this stage.

27

Protocol 1. *Continued*

4. Edit the sequence to remove secondary structure. Each of the 60 base sequences should be edited to remove self-complementary regions (see Section 2.2.2), if possible by trimming so that one is left with a sequence of 32–36 bases which has a GC content of about 50%. The finished sequences can be copied and pasted into the text box on the Genbank and a *BLAST* search performed, as described above.

5. Choose the best sequence or sequences. Typically there will be one or two sequences which are clearly preferable by having a lack of similarity to other genes in the Genbank sequence, a GC content closest to 50%, and relatively few regions of self-complementarity. Making two oligonucleotides rather than one is useful for control experiments (see below), and one oligonucleotide may outperform another, even though they appear similar.

6. Work out the reverse complement of the sequence(s).

[a] mRNA molecules also contain 5′ and 3′ untranslated regions which can be used as targets for oligonucleotide probes, but we generally avoid these regions for probe design since they are more likely to contain areas of high secondary structure.

2.2.2 Editing out self-complementary regions

Most commercial DNA software packages will allow analysis of self-complementarity and therefore allow a longer sequence to be edited to a short one with few stacking sites. Unfortunately however, at the time of writing, we know of no Internet-based programs which are able to help in the editing of larger sequences to remove hairpin loops and stacking sites. However, there are a number of Internet-based, freeware, shareware, and commercial computer programs which enable design of oligonucleotide primers for the polymerase chain reaction (PCR). Since many of the issues for PCR primer design are similar to that of oligonucleotide design for *in situ* hybridization, some of these programs might prove suitable, although they have not been fully evaluated by us. Important features when using a program are that it must allow the user to choose the length of the oligonucleotide (PCR primers are shorter than oligonucleotides for *in situ* hybridization), must be able to choose the GC content range, and specify the maximum length of self-complementarity that is acceptable.

We use the *GCG* package (4), which in the UK can be accessed for a small cost through the Human Genome Mapping Project Resource Centre (http://www.hgmp.mrc.ac.uk/). The *stemloop* program in the *GCG* package is excellent at determining stacking sites and sites for potential hairpin loop formation. Using this information, it is possible to edit the sequence (with the *seqed* program) to remove the worst regions of self-complementarity. Using these two programs through a number of cycles, it is usually possible to find an oligonucleotide of 32–36 base pairs of 45–55% GC content with a small number of stacking sites. These programs are not user-friendly as they require a degree of familiarity with Unix, and a complete description of them is beyond the scope of this chapter.

3. Oligonucleotide probe synthesis

3.1 Manufacturing oligonucleotides

There are many companies that synthesize oligonucleotides at very competitive prices and most will supply the oligonucleotides 48–72 h after ordering. Synthesis of an oligonucleotide for a laboratory, therefore, is as simple as writing down the sequence and sending it to your favourite manufacturer. However, it is very important to double check your sequence before it is sent off:

(a) Is the sequence 'antisense' and written in the 5′ to 3′ direction? The DNA strand shown in published sequences and in gene databases is always the coding or plus strand, and is written from left to right in the 5′ to 3′ orientation. Since this strand has a sequence homologous to the RNA which it encodes, the oligonucleotide probe therefore must be the reverse complement of the DNA strand (see Chapter 1).

(b) Have there been any errors in writing down the sequence?

(c) Have you written down the sequence in a legible way? Cs and Gs can be muddled up.

Oligonucleotides are synthesized chemically using an automated synthesizer on small columns (6). The synthesis proceeds in cycles, after which the oligonucleotide is cleaved from the column and desalted. Manufacturers offer a number of scales of oligonucleotide synthesis, typically 0.05, 0.2, and 1 micromole. Because each reaction cycle is not 100% efficient, the actual yield is lower than the theoretical amount, and purification of oligonucleotides also reduces the total amount of oligonucleotide. These factors can reduce the amount of oligonucleotide obtained to about 5–10% of the starting material. Since the amount of oligonucleotide used for one *in situ* hybridization experiment is 4 picomoles, even the smallest scale synthesis will yield enough for tens of thousands of experiments.

3.2 Oligonucleotide purification

During oligonucleotide synthesis, each cycle is not 100% efficient and therefore every oligonucleotide contains some 'failure sequences'—oligonucleotides which are shorter than the expected length. Failure sequences can make up to 30% of a crude oligonucleotide preparation. Often this will not be a problem, but failure sequences can cause increased non-specific hybridization. It is therefore useful to purify the oligonucleotide before use. Techniques to purify oligonucleotides include polyacrylamide gel electrophoresis (PAGE; *Protocol 2*) and HPLC-based methods (3). The PAGE method is considered to be the best, and most oligonucleotide manufacturers will carry this out at an extra cost.

Protocol 2. Polyacrylamide gel purification of oligonucleotides

Equipment and reagents

- Vertical gel apparatus: those that support a large area gel such as a sequencing gel are best (Gibco BRL model S2 sequencer which has a gel 38.5 cm in height), but a smaller gel frame such as those used for Western blots will do
- Glass plates and 1.6 mm width spacers
- 1.6 mm width combs to give several large wells
- Gel sealing tape (Gibco BRL, 11032–018)
- Power supply
- Hand-held transilluminator (short wave ultraviolet)
- Goggles or face mask for face protection from ultraviolet light
- Bench vacuum
- Fluorescent thin-layer chromatography plates (Sigma, T-6667)
- Glass rods

- Saran wrap or clingfilm
- 15 ml sterile tubes
- Bench centrifuge
- Sterile Eppendorf tubes
- 40% acrylamide:bisacrylamide solution (19:1)[a] (National Diagnostics, Accugel, EC-850)
- TEMED (Gibco BRL, 15524–010)
- Ammonium persulfate (Gibco BRL, 15523–012)
- Acetone
- Ultrapure water
- Xylene cyanol: 5% stock solution in water, stable for months–years
- Bromophenol blue: 5% stock solution in water, stable for months–years
- TBE: see *Table 1*
- Deionized formamide: see *Table 1*

A. *Setting-up the gel*

1. Mix components for 12% polyacrylamide gel: 120 ml 40% acrylamide: bisacrylamide solution,[a] 40 ml 10 × TBE, and make up to 400 ml with ultrapure H_2O.

2. Transfer solution to a very clean side-arm vacuum flask[b] and apply a vacuum for 20 min to degas the mixture.

3. Whilst the mixture is degassing, prepare the gel plates. Wearing gloves, wash plates well in soapy water and scrape off any fragments of old gel. Rinse plates twice in ultrapure water, with acetone, and rinse again with water. Dry with tissue. Assemble gel with side and bottom spacers, and seal with tape following manufacturer's instructions.

4. Add 0.2 ml of a freshly made 10% (w/v) ammonium persulfate solution and 50 μl TEMED. Gently swirl to mix, but take care not to aerate the mixture. Pour solution into the gel frame, and insert the comb.

5. When gel has set (about 1 h), carefully remove comb. Put into gel frame and add 1 × TBE buffer to top and bottom chambers. Using a syringe and needle, remove air bubbles from bottom part of gel and flush out wells with 1 × TBE.

B. *Running the gel*

1. Electrophorese the gel for 30 min at 20–25 V/cm to heat it up to approx. 60°C. Periodically check that buffer does not leak from the top chamber.

2. Whilst gel is pre-heating, take oligonucleotides and suspend in ultrapure H_2O to a concentration of approx. 5 nmol/μl.

3. Use approx. 100 nmol of your oligonucleotide (20 μl) and add an equal volume of deionized formamide (20 μl) and 0.5 vol. of glycerol (10 μl). Do not add dye to these samples. Make up two samples of dye (20 μl formamide, 10 μl glycerol, 2 μl bromophenol blue, 2 μl xylene cyanol).

4. Heat all tubes to 65°C for 15 min, then quench on ice.

5. Turn off power supply to gel. Flush wells with buffer using a syringe and load samples. Load a sample of dye at each end. Since the oligonucleotide samples do not contain dye, it can be difficult to see them. However close observation should allow visualization of the dense formamide solution entering the well. As a precaution, it is best to leave empty wells between oligonucleotides to minimize the effect of spillover.

6. Electrophorese the samples until xylene cyanol band (light blue) is about one-half to two-thirds of the way down the gel. Turn off power and remove top gel plate. Cover gel with clingfilm or Saran wrap. Place a fluorescent thin-layer chromatography plate face down on the clingfilm. Invert, then remove the other plate.

C. *Purification of oligonucleotides using ultraviolet (UV) shadowing*

1. In the dark, using a hand-held transilluminator (short wave UV),[c] visualize the oligonucleotide bands. The bands are visible since the oligonucleotides absorb UV light and therefore prevent the plate from fluorescing in the region underlying the oligonucleotide. There should be a single band. If the oligonucleotide contains many failure sequences, a ladder of bands of lower molecular weight will also be visible. Sometimes faint bands of high molecular weight are visible that are primer multimers.

2. Using a razor blade, trace around the outline of the oligonucleotide bands. Carry out this step as quickly as possible, since prolonged exposure to short wave ultraviolet light will cross-link oligonucleotides.

3. In the light, remove blocks of the gel containing the bands. Put into separate sterile tubes (15 ml) and crush gel using a clean glass rod. Add 1 ml ultrapure water.

Protocol 2. *Continued*

4. Heat mixture at 65°C for 2 h. Centrifuge tubes in a bench centrifuge at 1000 r.p.m. to sediment the gel fragments. Remove solution containing oligonucleotide to sterile Eppendorf tubes. This is the oligonucleotide working stock solution.

[a] Safety note: unpolymerized acrylamide is a neurotoxin. Wear protective clothing at all times, including gloves. Protect laboratory benches from spills by pouring gel on benches covered with tissue or a disposable coating. Any remaining acrylamide solution should be polymerized by addition of ammonium persulfate and TEMED, and the solid waste should be disposed of by incineration.
[b] Any traces of previous gels will produce rapid polymerization.
[c] Safety note: wear appropriate face protection (UV blocking face mask or goggles) for this work.

3.3 Quantification of oligonucleotide

For many molecular biological procedures DNA or RNA is quantitated as weights rather than absolute amounts (moles). However, for oligonucleotides it is important to know the molarity of your oligonucleotide since, when labelling, the molar ratio of label to probe should be known. A spectrophotometer is used to quantify oligonucleotides. It is important to use a very clean quartz cuvette to measure the absorbance at 260 nm. This absorbance is proportional to the amount of oligonucleotide and can be converted using a simple formula to obtain the molarity of the solution (formula derived from data in *Table 1*). A formula using a semi-micro quartz cuvette with a 1 cm path length is given in *Protocol 3*, but any good quality spectrophotometer will do, including small bench-top spectrophotometers designed to measure DNA concentration.

Table 1. Data for oligonucleotides

Molecular weight (M_r):	$(A \times 312) + (G \times 328.2) + (C \times 288.2) + (T \times 303.2) - 61$
1 OD_{260} unit:	30 μg of an oligonucleotide
nmol:	$1000/M_r \times$ μg of an oligonucleotide

Protocol 3. Quantification of oligonucleotides

Equipment

• Ultraviolet spectrophotometer
• Quartz semi-micro cuvette (1 cm path length)

Method

1. Clean the cuvette very well with ethanol and several rinses of ultrapure water.

2. In a sterile Eppendorf tube, pipette *x* μl (usually 5) of the oligo-nucleotide working stock solution.[a] Add 995 μl of ultrapure water. In a separate tube, take 1 ml of the same batch of ultrapure water as a blank.

3. Use the blank to set the absorbance at 260 nm to zero. Measure the absorbance of the sample at 260 nm.

4. Calculate concentration of oligonucleotide using the following formula (derived from *Table 1*): concentration (pmol/μl) = $(A_{260}/30) \times (1000/M_r) \times (1000/x)$.

5. Dilute oligonucleotide to a concentration of 1 pmol/μl in sterile water. Store in aliquots frozen at −70°C. Oligonucleotides are stable for months or years and can withstand a large number of freeze–thaw cycles.

[a] This applies for a PAGE purified oligonucleotide. For oligonucleotides from a manufacturer first dissolve in a small volume (e.g. 0.2 micromoles in 50 μl) and dilute an aliquot (say 5 μl) into 1 ml to make a working stock solution.

4. Oligonucleotide labelling

4.1 Choice of labelling method

Oligonucleotides can be labelled by a variety of methods. Manufacturers sell oligonucleotides with a number of chemical modifications including addition of biotin, fluorescein, and proteins (e.g. alkaline phosphatase) (7). These modifications can be very expensive so it is more common to purchase unmodified oligonucleotides and label them enzymatically.

Unmodified oligonucleotides have potentially reactive hydroxyl groups at the 5′ end and the 3′ end. The 5′ hydroxyl group can be phosphorylated using the enzyme T4 polynucleotidyl kinase, and it is therefore possible to radio-label oligonucleotides at this position using [γ-^{32}P]ATP, [γ-^{33}P]ATP, or [^{35}S]ATPγS as a phosphate donor (1). Under these conditions, each labelled oligonucleotide has a single radioactive label. More common is labelling of the 3′ end of an oligonucleotide probe with terminal transferase (TdT) (3). This enzyme adds deoxynucleotides sequentially to the oligonucleotide and, depending on the conditions used, can add four to ten labelled bases (or more) onto the end. A wide variety of modified deoxynucleotides can be used which include for radioactive probe labelling, [^{35}S]dATP, [^{32}P]dATP, [^{33}P]dATP, or [^{3}H]dATP. There are also an increasing number of modified nucleotides available for non-radioactive probe labelling including fluor-escein, biotin, digoxigenin, rhodamine, and coumarin, most of which, with the exception of digoxigenin (8) have not been widely used for detection of mRNA by *in situ* hybridization. In addition, it is possible to purchase nucleotides that have been modified to have amine groups that can be reactively coupled to proteins such as alkaline phosphatase.

The choice of labelling method is of central importance and each method has its advantages and disadvantages (see Chapter 1). One key decision is between non-radioactive and radioactive methods. Advantages of the radioactive methods are that they are very sensitive, and since it is easy to monitor the radiolabelling the investigator can be sure of using well-labelled material. The main disadvantages of radioactive methods are the need for special precautions for the use, handling, and safe disposal of radionucleo-tides, and that the label must be detected using autoradiographic methods which yield lower resolution than a non-radioactive method. Non-radioactive methods have the great advantages of a high resolution and of yielding results more quickly than radioactive-based methods. However, it is often difficult to monitor the degree of labelling of the probe, such that experiments could be carried out using a poorly-labelled probe which yields no useful results. In addition, we find that most non-radioactive oligonucleotide probes are less sensitive than radioactive methods. This could be due to a number of factors including that the modified bases are bulky and may reduce specific hybridiza-tion by steric hindrance. In addition, depending on the modified base used, the 3' tail might be shorter than for radiolabelled probes since these bulky nucleotide analogues may inhibit or have a lower affinity for terminal trans-ferase. We recommend therefore that when novel oligonucleotide probes are used, radioactive rather than non-radioactive labelling is used.

4.2 3' end-labelling with radiolabelled nucleotides

4.2.1 General considerations

The main factors that govern the choice of radiolabel are: the energy of the β-emitter used, the mean free path length of the emission, the time it takes for the isotope to decay (half-life), and its cost (*Table 2*). The higher the energy of the β-emitter, the shorter the exposure time that will be necessary. The path length relates to the distance that the β-particles (electrons) will travel away from the source of hybridization before collision with a silver ion to produce a deposition of silver grains. A large path length yields a low resolution. In practice this means that the highly energetic ^{32}P allows short exposure times but the high path length means that the signal is of low resolution, and may not even be suitable for film autoradiography. Conversely the low energy ^{3}H requires exceptionally long exposures and is therefore only useful where very high resolution is required, for example in subcellular localization of RNAs. ^{35}S and ^{33}P have a moderate energy of emission allowing exposure times of days for film autoradiography and weeks for liquid emulsion autoradiography. The path lengths yield a resolution of 10–15 μm and 15–20 μm for ^{35}S and ^{33}P probes, respectively, suitable for localization of mRNAs in single cells using liquid emulsion techniques. Because of its longer half-life (*Table 2*), ^{35}S-

Table 2. Physical properties of radiolabelled nucleotides

Isotope:	[32]P	[33]P	[35]S	[3]H
Energy of emission (MeV)	1.709	0.249	0.167	0.019
Path length in water	0.8 mm	0.6 mm	0.32 mm	0.006 mm
Half-life	14.3 days	25.4 days	87.4 days	12.4 years

labelled probes can be stored before use longer than ^{33}P-labelled probes, and if autotoradiographic exposure times of greater than one month are necessary, the former probes will give greater signal strength. In addition, at the time of writing, ^{33}P-labelled nucleotides are over twice the cost of ^{35}S-labelled nucleotides. ^{35}S-labelling is thus our method of choice.

4.2.2 End-labelling with [^{35}S]dATP

A protocol for labelling with [^{35}S]dATP is given in *Protocol 4*.

Protocol 4. 3′ end-labelling of oligonucleotides with [^{35}S]dATP

Equipment and reagents

- Water-bath or heating block set to 37°C
- Microcentrifuge
- [^{35}S]dATP (deoxyadenosine 5′(α thio) triphosphate) at 1000 Ci/mmol, 10 μM. Amersham (grade for 3′ end-labelling, SJ1334) or NEN (NEG034H) are suitable. Isotopes with a high concentration of dithiothreitol will cause problems through precipitation of cobalt ions from TdT buffer. To avoid frequent freeze–thawing, store isotope in 3 μl aliquots at –70°C for up to five weeks before use.
- Sterile microcentrifuge tubes
- Sterile pipette tips
- Eppendorf pipettes or equivalent
- Radiation safety equipment
- Oligonucleotide (1 pmol/μl)
- Ultrapure water (autoclaved): do not use DEPC.H$_2$O
- TdT enzyme (terminal deoxynucleotidyl transferase) from Promega, Amersham, Gibco BRL, or MBI Fermentas: use a make that is supplied with tailing buffer since this is difficult to prepare
- 5 × tailing buffer

Method

1. In a sterile microcentrifuge tube, add the following components in the order shown to give a reaction volume of 50 μl:

   - oligonucleotide (4 pmol)   4 μl
   - 5 × tailing buffer   10 μl
   - ultrapure water   27 μl
   - BSA (2 mg/ml)   1 μl
   - [^{35}S]dATP   3 μl
   - TdT (40–60 U)   5 μl

2. Seal tube well, flick tube to mix, and spin for 10 sec in a micro-centrifuge.

Protocol 4. *Continued*

3. Incubate reaction for 1 h at 37 °C.

4. Stop reaction by freezing on dry ice, or proceed directly to purification of labelled oligonucleotide (*Protocol 5*).

4.2.3 Purification of labelled probes

After radiolabelling a probe, it is important to remove unincorporated nucleotides. These nucleotides can give rise to high non-specific backgrounds since they may bind to positively charged groups on tissue sections. It is possible to remove unincorporated nucleotides by ethanol precipitation or size exclusion chromatography (1). Spin columns for size exclusion chromatography can be used, but we have found that the yields of labelled oligonucleotide are lower than if larger columns are used (*Protocol 5*).

Protocol 5. Purification of ^{35}S-labelled oligonucleotides using Sephadex columns

Equipment and reagents

- Sephadex G50 columns (Nick columns, Pharmacia)
- Sterile disposable test-tubes (15 ml)
- Sterile pipette tips
- Eppendorf pipettes or equivalent
- Test-tube rack
- Parafilm
- Radiation safety equipment
- Hand-held Geiger counter

- Freeze drier *or* vacuum centrifuge *or* microcentrifuge
- TE (10 mM Tris, 1 mM EDTA): make up from stock solutions (*Table 1*) of 1 M Tris pH 8 and 0.5 M EDTA pH 8 by diluting into DEPC.H$_2$O; autoclave before use
- 1 M DTT (dithiothreitol) stock (*Table 1*): for 20 mM DTT solution dilute the stock with DEPC.H$_2$O; store in 100 μl aliquots in sterile Eppendorf tubes at –20 °C

Method

1. Wrap Parafilm around Nick column near the bottom and stand the column over the test-tube.

2. Remove cap (at top) and the stopper (at bottom) of the Nick column and add 4 ml TE to the column. Allow solution to drip into the collection tube.

3. When the all the solution has dripped through the column, stopper the column. Do not allow the column to dry.

4. Add labelled probe to the top of the column. Remove stopper from column.

5. Once probe solution has disappeared add 400 μl TE to top of the column.

6. When all of the solution has dripped through the column, stopper the column. Transfer column to a new labelled test-tube which contains 5.5 μl 1 M DTT.

7. Add 550 μl TE to top of the column and remove stopper from column. Collect effluent which contains the labelled probe.

8. Take 5.5 μl of probe for scintillation counting (*Protocol 6*A), or scan effluent and column with hand-held Geiger counter (*Protocol 6*B) to determine probe specific activity.

9. Concentrate probe in a freeze drier, a vacuum centrifuge, or by ethanol precipitation.[a] Resuspend in 20 μl 20 mM DTT and store frozen at –70 °C.

[a] To ethanol precipite the sample add 55 μl 3 M sodium acetate pH 5.2 (*Table 1*), 2 μl glycogen (10 mg/ml, Boehringer Mannheim, 901 393), and 2 ml ethanol. Divide the solution between two microcentrifuge tubes, freeze at –70 °C or on dry ice for at least 30 min, or at –20 °C overnight. Spin at 15 000 *g* for 30 min and pour off the ethanol. Invert the tubes to drain off the ethanol and air dry for 30–60 min before resuspending pellets.

4.2.4 Specific activity of labelled probes

It is essential to determine how well the probe is labelled. *Protocol 4* uses a 1:7.5 molar ratio of oligonucleotide:labelled nucleotide, which should result in seven to eight nucleotides being added to the 3′ end. Since each [^{35}S]dATP molecule has a specific activity of approximately 1000 Ci/mmol, the specific activity of the labelled probe should therefore be 7500 Ci/mmol. In practice the specific activity will vary—the reaction might not work with optimum efficiency or there may be pipetting errors which will affect the relative concentrations of oligonucleotide and labelled nucleotide. Also, the actual specific activity of the probe will depend on the amount of radioactive decay, i.e. the days before or after the activity date of the batch of isotope. We have found that for ^{35}S-labelled oligonucleotides it is not worth carrying out experiments unless the probe has a specific activity between 4000–10 000 Ci/mmol. Tails shorter than four will produce a very low signal strength, whereas tails greater than ten can give problems with non-specific hybridization (9). Two methods of estimating the specific activity of probes are given in *Protocol 6*.

Protocol 6.   Calculation of specific activity of ^{35}S-labelled probes

Equipment and reagents

- Scintillation counter
- Hand-held Geiger counter that is sensitive to β-radiation
- Clean scintillation vials with caps
- Scintillation fluid, e.g. Ecoscint A (National Diagnostics, LS-273)

A. *Accurate method*

1. Pipette 1% (v/v) (e.g. 5.5 μl from 550 μl) of the labelled probe into a scintillation vial. Add required amount of scintillation fluid, stopper, and shake well.

2. Count vials in a scintillation counter, using a programme set for 1 min counts for ^{35}S.

Protocol 6. *Continued*

3. Convert the counts per minute (c.p.m.) to disintegrations per minute (d.p.m.).[a]

4. Multiply by 100 to determine the total d.p.m. in your sample.

5. Convert to Curies (Ci): divide by 2.2×10^{12}.

6. Calculate specific activity in Ci/mmol.[b]

B. *Quick method*

1. After running the reaction mixture through a Nick column, use a hand-held Geiger counter to estimate the counts per second (c.p.s.) from the collection tube (probe counts) and the c.p.s. remaining in the column (column counts).

2. Calculate the proportion of total counts in probe = probe counts / (probe counts + column counts).

3. Multiply this figure by 7500. This will give you an estimate of the probe's specific activity in Ci/mmol.

---

[a] Most modern scintillation counters will do this automatically. If not make accurate serial dilutions of [^{35}S]dATP to estimate the counting efficiency of the counter (1 Ci = 37 MBq = 2.2×10^{12} d.p.m. = [2.2×10^{12} c.p.m. ÷ counting efficiency of counter]).
[b] If 4 pmol of an oligonucleotide was labelled, divide by 4×10^{-9}.

4.3 Non-radioactive 3′ end-labelling

We have successfully used fluorescein labelled and a digoxigenin labelled nucleotide to label oligonucleotide probes. Biotinylated oligonucleotides probes have been used by a number of groups (10), although there have been problems reported with high non-specific signal under some conditions (11). In our hands, fluorescein labelled oligonucleotides work better than digoxigenin oligonucleotides, and we provide protocols for their labelling (*Protocol 7*) and use (*Protocol 14*). However, the results we get with non-radioactive probes have almost always been inferior to ^{35}S-labelled probes.

Protocol 7. 3′ end-labelling with a fluorescein-modified nucleotide

Equipment and reagents

- Water-bath set to 37 °C
- Sterile microcentrifuge tubes
- Sterile pipette tips
- Eppendorf pipettes or equivalent
- Microcentrifuge
- Positively charged nylon membrane (Hybond N$^+$, Amersham)
- Oligonucleotide (10 pmol/μl)
- Ultrapure water (autoclaved): do not use DEPC.H$_2$O

- Fluorescein-N^6-dATP (DuPont/NEN, NEL501)
- Terminal deoxynucleotidyl transferase (TdT)
- 5 × tailing buffer (supplied with TdT)
- Ultraviolet light transilluminator
- Ultraviolet light safety equipment (face mask)
- 100% ethanol
- 20 × SSC (see *Table 1*)

A. *Setting-up the reaction*

1. In a sterile microcentrifuge tube combine the following to reach a total volume of 40 μl:

   - oligonucleotide (25 pmol)   2.5 μl
   - 5 × tailing buffer   8 μl
   - ultrapure water   23 μl
   - fluorescein-dATP (25 nmol)   2.5 μl
   - TdT (32 U)   4 μl

2. Flick test-tube to mix and centrifuge the solution to the bottom of the tube in a microcentrifuge (approx. 10 sec).

3. Incubate reaction for 1–2 h at 37 °C.

4. No further purification is usually necessary. Reaction product can be stored at −70 °C until use.

B. *Determining degree of labelling of probe*

1. Wearing gloves, cut two small squares of Hybond N^+ membrane (approx. 2 cm × 3 cm). Label filters using a pencil.

2. In Eppendorf tubes, make a serial dilution of fluorescein-N^6-dATP in DEPC.H_2O (1/100, 1/300, 1/1000, 1/3000, 1/10000; these can be stored at −20 °C for reuse). Spot 1 μl of each of these dilutions on each filter.

3. Spot 1 μl of labelled probe onto each filter.

4. Wash one filter for 15 min at 60 °C in 2 × SSC, 0.1% SDS (w/v). This procedure will wash off any fluorescein labelled nucleotide that has not been incorporated into the oligonucleotide.

5. Rinse filters briefly in cold water and then in ethanol.

6. Wrap in clingfilm or Saran wrap and view on a transilluminator. Comparison of the filters should show that the fluorescein-dATP standards are not visible on filters that have been washed, but the probe spot is still visible. A well-labelled probe will have a fluorescence intensity approximately the same as the 1/300–1/1000 dilution standard on the filter that has not been washed.

5. *In situ* hybridization with oligonucleotides

5.1 Tissue preparation

In situ hybridization can be performed on tissues prepared in a variety of different ways. For ^{35}S-labelled oligonucleotides we have consistently found the best results are obtained using sections that are prepared from frozen tissue. It is important to freeze tissue as quickly as possible after removal since RNA is unstable (12). It is possible to use sections that are completely unfixed (13), but we prefer to use sections that are cut onto slides then fixed lightly, acetylated, delipidated, and dehydrated (*Protocol 8*). Sections prepared from animals that have been carefully perfused also work well (*Protocol 9*), but can give a higher background. For fluorescein labelled oligonucleotides, we have most successfully used sections from perfused animals which are cut and left free-floating during the incubation procedures (*Protocol 13*). We also routinely use perfused material for combined immunocytochemistry with *in situ* hybridization (*Protocol 14*).

It is also possible to use oligonucleotides on sections cut from tissues that are preserved in formalin or embedded in wax or paraffin (14). We have had less experience and success in these methods. To some extent this may be due to the quality of RNA remaining in the tissue, and often material prepared for immunocytochemical procedures has been treated in a way which is less than optimal for preservation of RNA. In addition, it may be necessary to include additional steps in the fixation/pre-hybridization protocol including proteinase K digestion and treatment with detergents to improve oligonucleotide penetration. We have not included protocols for these steps in this chapter (see Chapters 3 and 4).

5.1.1 Subbed microscope slides

Slides utilized for mounting sections must be 'subbed' before use to allow sections to attach and prevent them falling off during processing. Subbing may also minimize non-specific binding of probes to glass slides. Slides may be coated with traditional agents such as gelatin chrome-alum, but this does not provide adequate adherence for all tissues. We recommend poly-lysine coated slides (*Protocol 8*A), or positively charged slides (e.g. Superfrost Plus slides, BDH). We have experienced problems with Superfrost Plus slides showing increased backgrounds when sections have been pre-treated and then stored at –70°C before use. Such storage is not always necessary for *in situ* hybridization combined with immunocytochemistry.

5.1.2 Use of cell culture for *in situ* hybridization

Cultured cells are suitable for *in situ* hybridization, provided they are grown on microscope slides or coverslips that can be glued onto slides. If the cells are grown on glass coverslips, then the coverslips should be glued onto slides as indicated in *Protocol 8*. A glass glue (readily available from DIY shops) should be used since other glues do not withstand the hybridization and post-hybridization procedures.

Protocol 8. Preparation of fresh frozen tissue sections or cells for *in situ* hybridization

Equipment and reagents

- Clean plastic staining dishes (BDH, 406/0236/00) or glass staining dishes
- Clean plastic slide racks (BDH, 406/0237/00): sterilization of racks and dishes can be carried out by soaking in 1 M NaOH for 1 h or 1% hydrogen peroxide overnight, followed by several washes in DEPC.H₂O
- Slide boxes (BDH, 406/0286/00)
- Magnetic stirrer hot plate
- Pre-cleaned microscope slides with frosted ends
- Poly-L-lysine solution, 100 μg/ml (Sigma, P8920)
- Sterile 15 ml tubes
- Acetic anhydride

- DEPC treated PBS, 0.9% NaCl, H₂O (see *Table 1*)
- 4% paraformaldehyde. Make on the day of use. In fume-hood weigh 10 g paraformaldehyde powder into 1000 ml glass beaker, add 250 ml DEPC.PBS, stir, and heat to about 70°C to dissolve. Cool and if necessary adjust to pH 7.4 with 1 M sodium hydroxide. Cool on ice before use.
- 0.1 M triethanalomine in DEPC-0.9% NaCl: stir 3.73 g triethanolamine in 250 ml DEPC-0.9% NaCl solution for 10–15 min until dissolved
- Chloroform

A. Preparation of poly-lysine coated slides

1. In a sterile tube, prepare a solution of 10 μg/ml poly-lysine in DEPC.H₂O.

2. Using a plastic Pasteur pipette, place one or two drops of poly-lysine on each slide.

3. Using a clean slide, draw the poly-lysine across the slide's surface so that the slide is coated.

4. Cover slides to prevent dust settling on them and leave for about 1 h to dry.

5. Store slides in boxes at 4°C until use.

B. Section or cell preparation on slides

1. For fresh frozen tissue: cut 10 μm sections using a cryostat onto poly-lysine coated slides. Store slides in boxes on ice during cutting. After cutting sections, leave slides at 4°C (in a fridge or on ice) for 1 h before pre-treatment.

2. For cells: grow cells on sterile microscope slides coated with a suitable substrate (e.g. collagen, poly-lysine). If slides are not convenient, grow cells on coverslips and glue the coverslips to microscope slides using a glass glue. Wash cells twice with PBS or serum-free medium.

C. Pre-hybridization

Place slides in plastic racks and carry out the following steps in staining dishes that hold 250 ml of each solution.

1. 15 min in 4% paraformaldehyde (use 2% paraformaldehyde for cultured cells).

Protocol 8. *Continued*

2. Wash twice (10 min each) in DEPC.PBS.

3. Acetylate: 10 min in 0.1 M triethanolamine in DEPC-0.9% NaCl, to which 0.625 ml acetic anhydride is added (final concentration 25 mM). Add acetic anhydride about 30 sec before putting the slide racks in, and stir vigorously to mix. Use once only per rack of slides.

4. Dehydrate through 70%, 95%, 100% ethanol, 5 min each. Ethanol solutions can be filtered and stored in tightly capped bottles for reuse.

5. Delipidate in chloroform for 5 min.

6. Remove chloroform by placing in 95% ethanol for 5 min.

7. Place slides on clean tissue, cover so that dust does not fall onto slides, and air dry for up to 1 h.

8. Either proceed directly to *in situ* hybridization (*Protocols 11* and *12* or *14*), or load into slide boxes and store frozen at –70 °C.

Protocol 9. Preparation of tissue by perfusion fixation

Equipment and reagents

- Perfusion apparatus[a]
- Fume-cupboard suitable for carrying out perfusion
- Dissection tray
- Dissection instruments including fine and standard size point-ended scissors and forceps, haemostats—metal and plastic (polypropylene forceps, BDH, 406/0058/00)
- Tissue embedding medium (OCT compound, Agar Scientific, R1180)
- Fine artist's paintbrush
- Hot plate magnetic stirrer
- pH meter or sensitive pH papers (Fluka Pehanon)
- Subbed microscope slides
- Clean glass vials (Agar Scientific, B794)
- Sodium pentobarbitol
- Saline solution: 0.9% NaCl in ultrapure H_2O
- Sucrose/ethylene glycol cryoprotectant solution (if required for storage of sections): dissolve 120 g sucrose, 120 ml ethylene glycol in 200 ml DEPC.PBS—adjust volume to 400 ml with DEPC.PBS
- DEPC treated 0.2 M and 0.1 M phosphate buffer, PBS, and H_2O (*Table 1*)
- 4% paraformaldehyde. This fixative must be made fresh on the day it is to be used. For 1000 ml, in a fume-cupboard add 40 g paraformaldehyde to 500 ml ultrapure H_2O and warm to 58–60 °C on a hot plate magnetic stirrer. Add several drops of 1 M NaOH until solution clears. Do not heat for extended periods or exceed this temperature as this adversely affects the properties of the fixative. Add 500 ml 0.2 M phosphate buffer and cool to room temperature. Check the pH, adjusting with NaOH or HCl if necessary
- 15–20% sucrose in 0.1 M phosphate buffer: in an oven baked glass beaker, weigh out 15–20 g sucrose. In a separate beaker, mix 0.2 M phosphate buffer with an equal volume of DEPC.H_2O. Add 80 ml to the sucrose and stir until dissolved, using a magnetic stir bar that has been autoclaved. Make up to 100 ml and store in aliquots in sterile 15 ml tubes at 4 °C. Do not autoclave

A. Perfusion

1. Fill the drip bottles with saline and 4% paraformaldehyde and hang them above the dissection tray in the fume-cupboard such that a good flow rate can be achieved. A pressure of between 80–120 cm H_2O is

required (normal rat blood pressure = 120 cm H_2O, 90 mm Hg). Allow sufficient amounts of each solution to flow through the tubing to remove all air bubbles from within it. Rinse out any fixative from the last portion of the tubing with saline solution.

2. Deeply anaesthetize the animal. For a rat, between 0.1–0.2 ml of a 60 mg/ml solution of sodium pentobarbital per 100 g weight is sufficient. When no longer responsive, place the animal on its back on the dissection tray.

3. Using forceps to hold the skin and muscle of the upper abdomen, make a mid-line cut just below the xiphoid process. Extend this incision laterally below the rib cage revealing the still intact diaphragm. From this stage onwards work as quickly as possible.

4. Holding the xiphoid process with forceps, cut along the top edge of the diaphragm into the thoracic cavity and make cuts laterally through the rib cage to form a flap of tissue which may be cut away or held back with a haemostat. There should be clear access to the heart and visibility of the ascending aorta.

5. Holding the apex of the heart with forceps, use fine scissors to cut the right atrium. Then make a small incision in the left ventricle and insert the perfusion cannula into the ventricle and carefully up into the ascending aorta until it is visible through the vessel. Do not advance it too far into the aorta as this may occlude the opening of the cannula. The cannula can be clamped gently in place using polypropylene forceps.

6. Perfuse the rat with approx. 50–100 ml of sterile saline solution. Carefully start the flow of the solutions so that no air bubbles enter the vasculature. This should take less than 5 min. Tissues should begin to appear more pale relatively quickly and the flow of blood from the right atrium should clear as blood is replaced by saline. If the lungs begin to fill with saline, it is likely the cannula has entered the left atrium. The cannula must be repositioned in the aorta to obtain a good perfusion.

7. Carefully change the stopcock position to allow the fixative to flow into the animal. Allow approx. 300–400 ml of fixative to perfuse the tissues.

B. *Handling of perfusion fixed material, post-fixation, and cutting*

1. Paraformaldehyde inactivates ribonucleases and hence degradation of mRNAs is inhibited in the fixed tissue. Gloves should be worn and precautions taken as outlined in Section 7 to avoid RNase contamination from exogenous sources.

2. Dissect out the tissues required and post-fix in a vial containing fixative for a further 1.5–2 h at 4°C.

Protocol 9. *Continued*

3. After this stage, sections can be cut *or* frozen and then cut. The choice of method will depend on the equipment available and the type of probe and hybridization procedure to be used.[b]

   (a) Cutting perfusion fixed tissues using a vibratome. The specimen holder and buffer well should be treated with 1 M NaOH and rinsed with DEPC.H$_2$O to reduce nuclease activity. Tissue should be immersed in DEPC-0.1 M phosphate buffer or DEPC.PBS during cutting and 20–25 μm sections[b] cut and collected free-floating in clean glass vials containing DEPC.PBS. If sectioning cannot be done immediately after the post-fixation, the tissue can be transferred to DEPC.PBS and kept at 4°C for up to 24 h before cutting.

   (b) Cutting perfusion fixed frozen tissue using a microtome or cryostat. Tissue must be cryoprotected before freezing. After post-fixation, transfer the tissue to 15–20% sucrose in phosphate buffer and allow to equilibrate overnight at 4°C. Freeze cryoprotected tissue in a block of tissue embedding media (OCT compound) on crushed dry ice or in isopentane maintained at –40°C. Following freezing, depending on which *in situ* hybridization protocols are to be used,[b] *either:* using a cryostat, cut thin sections (5–15 μm) that are mounted directly onto subbed slides;[c] *or:* using a sliding microtome or cryostat, cut thicker sections (20–30 μm) from frozen tissues, collecting them free-floating in DEPC.PBS.

4. If desired, free-floating sections[b] can be mounted onto subbed microscope slides[c] using a fine artist's paintbrush.

5. It is recommended that sections cut from perfusion fixed tissue should be used immediately for *in situ* hybridization. If this is not possible, both free-floating and slide mounted sections can be stored in sucrose/ethylene glycol cryoprotectant solution at –20°C for several months up to one year.[d] After storage in this solution, sections must be washed at least five times for 10 min in DEPC.PBS before continuing with other protocols.

6. For slide mounted sections use *Protocols 8*C, *11*, and *12* for radiolabelled probes; *Protocols 8*C and *14* for fluorescein labelled probes. For free-floating sections see Section 5.2.3, and use *Protocols 13* and *14* for fluorescein labelled probes. Use *Protocol 15* for *in situ* hybridization combined with immunocytochemistry.

---

[a] Adequate pressure is necessary to rinse blood from the cardiovascular system of the animal (with saline) and ensure that fixative perfuses the tissues. We routinely utilize a gravity-based perfusion apparatus. This assembly consists of two 1.5 litre drip bottles (one for saline and one for fixative) attached to two lengths of clear flexible tubing joined by a three-way stopcock and a final piece of tubing. We use Keck ramp clamp tubing clamps (Sigma, Z22, 229–1) on each of the lengths of tubing to regulate flow rate of the solutions. At the very end of the tubing a

blunt-ended metal cannula approx. 2–4 cm in length and 1–2 mm in diameter (depending on the size of the animal to be perfused) is attached. Adaptations of this basic set-up may be necessary in some cases such as when perfusing small animals (e.g. mice or young rats) where a gravity feed might be replaced by syringes and manual pressure.

[b] Thick sections cut on a vibrotome or microtome are appropriate for non-radioactive *in situ* hybridization, which we recommend to be carried out on free-floating sections. For radioactive probes, thick sections when mounted onto slides can give high background labelling. Thinner sections cut using a cryostat directly onto slides are generally recommended for use with radioactive oligonucleotides. Section 5.2.2 discusses the use of radioactive probes on free-floating sections.

[c] Poly-lysine coated slides (*Protocol 8A*) or charged slides (Superfrost Plus, BDH) are recommended.

[d] In most cases antigenicity is maintained in sucrose/ethylene glycol cryoprotectant allowing subsequent immunocytochemistry. No degradation of mRNAs has been noted. However, some small antigens and certain tracers such as FluoroGold or Fast Blue can be lost with storage in cryoprotectant. In these cases, storage in cryoprotectant is not recommended.

5.2 Hybridization and post-hybridization washes

5.2.1 Protocols for ^{35}S-labelled oligonucleotides on slide mounted sections

Just before use, suspend the probe in a suitable volume of hybridization buffer (*Protocol 10*), calculated according to the number of slides needed. We use most radiolabelled probes at a concentration of 2 nM, e.g. 4 pmol (20 μl) of probe in 2 ml hybridization buffer, which is enough for 25–30 slides. Probes must be heated to remove any secondary structure before use, and then kept on ice until used. Hybridizations are carried out according to *Protocol 11*, and washing according to *Protocol 12*. If lower stringency conditions are required, for example if using very short oligonucleotides, AT-rich oligonucleotides, or carrying out cross-species hybridization, then reduce the washing temperature in steps 3 and 4. If higher stringency conditions are required, e.g. to eliminate hybridization to sequences mismatched by one or two bases, then increase the washing temperature in steps 3 and 4. Similarly, a reduction in salt concentration (SSC) will increase stringency.

Protocol 10. Preparation of hybridization buffer

Equipment and reagents

- 50 ml sterile tubes
- Clean weighing boats
- Oven baked spatulas for weighing out reagents
- Water-bath at 65°C
- Vortex mixer
- Dextran sulfate, sodium salt, $M_r \sim 500\,000$ (Pharmacia, 17–0340–01)

- Polyadenylitic acid (Sigma, P9403): add DEPC.H_2O to the bottle to give a concentration of 10 mg/ml—the solution will need to be heated and vortexed
- Denatured, sheared herring sperm DNA (10 mg/ml) Deionized formamide, 1 M DTT, DEPC treated H_2O, and 20 × SSC (see *Table 1*)

Hybridization buffer can be made up in 50 ml batches which are stable for several months at –20°C. Take care to minimize RNase contamination when preparing the stock solution (Section 7).

Protocol 10. *Continued*

Method

1. Weigh out or thaw the following reagents:

Volume used	Component	Final concentration
0.5 ml	100 × Denhart's solution	(1 ×)
10 ml	DEPC-20 × SSC	(4 ×)
250 μl	polyadenylitic acid	50 μg/ml
2.5 ml	denatured, sheared herring sperm DNA	500 μg/ml
10.75 ml	DEPC.H$_2$O	—
5 g	dextran sulfate	10%
25 ml	deionized formamide	50%

2. In 50 ml sterile tube, add dextran sulfate powder slowly to the first five ingredients.

3. Heat to 65°C and vortex frequently until dextran sulfate is completely dissolved. This will take 20–30 min.

4. Add formamide, mix well, store at –20°C. Before use, heat solution to 65°C and vortex to mix.

5. For use with ^{35}S-labelled probes add 1 M dithiothreiotol to give a final concentration of 10 mM.

Protocol 11. Hybridization of radiolabelled oligonucleotide probes to slide mounted sections

Equipment and reagents

- Plastic or polypropylene sandwich boxes with tightly-fitting lids
- Plastic staining dishes (BDH, 406/0235/00)
- Plastic slide racks (BDH, 406/0235/00)
- Filter paper (Whatman 3MM)
- Water-bath or heating block set to 65°C
- Incubator oven, or large water-bath with lid, set to 37°C
- Vortex mixer
- Large trays or boxes to maintain 'dust-free' area to dry slides
- Ice
- Glass coverslips
- Lint-free tissue
- Hybridization buffer (*Protocol 10*)
- 2 × SSC: 10 ml 20 × SSC plus 190 ml water
- 100% ethanol in a clean glass beaker
- Radiation safety equipment
- Blunt seeker or 20 gauge needle

A. *Setting-up*

1. Prepare slides: if using freshly pre-treated slides, continue with step 2. If sections have been stored frozen, remove slides from freezer and place in plastic rack. Using staining dishes, dehydrate in 50%, 70%, and 95% ethanol for 5 min each. Place slides on tissue and cover with an upside-down tray or box to dry in a 'dust-free' environment. Do not leave for longer than 1 h before hybridization.

2. Prepare RNase treated sections as controls (*Protocol 16*).

3. Prepare probe. Remove probe from freezer and dissolve probe in hybridization buffer to a concentration of 2 nM in a 15 ml sterile tube. Each slide requires 60–80 μl probe, but because the hybridization buffer is viscous, it is best to allow extra. Heat probe to 65°C for 15 min to remove secondary structure and immediately place on ice.

4. Prepare coverslips. Take glass coverslips and dip individually in fresh 100% ethanol. Wipe dry with lint-free tissue and leave to dry covered by tissue. Discard any coverslips that are broken, smeary, or have visible lint or dust on them.

5. Prepare humidified boxes. Only use boxes which have tightly-fitting lids; the best boxes are made of soft plastic. Cut a piece of filter paper (Whatman 3MM) to fit snugly on the bottom of each box. Add 10 ml 2 × SSC, and tilt box so that the filter paper is completely wet. Make sure that there are no wrinkles in the filter paper. Invert boxes to allow surplus solution to drain off before use.

6. Label slides on their frosted ends using a pencil.

B. *Hybridization*

1. Apply 80 μl of probe to one slide. Pipette this in two or three drops in the centre and at one end of the slide.

2. Using the corner of a coverslip, carefully remove any larger air bubbles.

3. Using a needle or seeker, lower the coverslip onto the slide so that the hybridization buffer covers the surface of the slide. Do not squash the coverslip onto the slide. If necessary hold the slide vertically to enhance the spread of the probe.

4. Repeat the procedure for the next slide. Do not be tempted to pipette probes onto multiple slides before applying coverslips since the hybridization buffer can dry very quickly and cause non-specific binding.

5. Load slides into humidified boxes. Make sure that each slide lies flat, and that slides do not overlap.

6. Fit lid to boxes, and check that the boxes are sealed by pressing on the top of each box.

7. Incubate at 37°C overnight, preferably in an incubator, but floating in a water-bath will do.

8. The following day, carry out post-hybridization washing (*Protocol 12*).

Protocol 12. Post-hybridization washing for radiolabelled oligonucleotide probes on slide mounted sections

Equipment and reagents

- Water-bath set to 55°C
- Plastic staining dishes (BDH, 406/0236/00)
- Plastic slide racks (BDH, 406/0237/00)
- Large trays or boxes to maintain 'dust-free' area to dry slides
- Radiation safety equipment
- Autoradiography cassette

- Autoradiographic film suitable for high sensitivity work (Amersham βmax Hyperfilm)
- 1 × SSC solution, room temperature (25 ml 2 × SSC plus 475 ml ultrapure H$_2$O) (see *Table 1*)
- 1 × SSC solution, pre-heated to 55°C
- Graded ethanol series: 50%, 70%, and 95% ethanol

Method

The following steps are carried out in plastic staining dishes. Care should be taken that the slides do not dry out during this procedure, particularly at the first stage.

1. Remove coverslips carefully.[a] Coverslips should ease off the slides.[b] Place slides in a rack in 1 × SSC.

2. Wash slides in 1 × SSC at room temperature for 10 min.

3. Wash slides in 1 × SSC at 55°C, twice for 30 min.

4. Wash slides in 1 × SSC at room temperature for 5 min.

5. Wash slides in 50% ethanol, then 70% ethanol, then 95% ethanol, for 15 min each step.

6. Leave to air dry in dust-free environment (10–60 min).

7. Load into an X-ray film cassette, add beta-sensitive film (βmax Hyperfilm, Amersham), and leave at 4°C for one week. After exposure, slides can be dipped in a photographic emulsion for high-resolution work (*Protocol 17*).

[a] Safety note. Most of the radioactivity applied to the slides is retained on the coverslip after removal, therefore carry out this step (and step 2) using appropriate radiation protection, and dispose of coverslips as radioactive solid waste. The solutions from step 1, step 2, and step 3 should be treated as radioactive liquid waste, but thereafter the amount of radioactivity to be found on the slides or in the liquids is minimal, and the slides can be handled safely without need for a shield.
[b] If they are stuck, it is likely that too little probe was added or that the humidified boxes were not properly sealed. These problems are likely to cause high non-specific binding of probe.

5.2.2 The use of free-floating sections from perfusion fixed tissues for *in situ* hybridization

Protocol 13 describes the pre-treatment of free-floating sections. We recommend the use of free-floating sections for fluorescein labelled probes, but, although feasible, not for *in situ* hybridization with radioactive oligonucleotides. Movement of sections from solutions can be somewhat more messy than transferring sections on slides and therefore extra care for radiation safety is advised. In addition, the increased tissue thickness necessary for cutting and handling of free-floating sections can result in unacceptable background levels of radioactivity on the sections.

Free-floating sections require special handling to maintain their integrity during processing. For this reason we only dehydrate the tissues briefly in 70% ethanol and do not use a delipidation step in the pre-treatments. Instead, we utilize a pre-hybridization step to ensure that the tissue is sufficiently equilibrated with hybridization buffer.

Protocol 13. Pre-treatment of free-floating sections with fluorescein labelled probes

Equipment and reagents

- Sterile Eppendorf tubes
- Small glass vials (Agar Scientific, B794) for washing sections and an artist's fine paintbrush for moving sections, both treated briefly in 1 M NaOH and rinsed with DEPC.H$_2$O
- DEPC.H$_2$O, DEPC.PBS (see *Table 1*)
- 0.1 M triethanolamine in DEPC.PBS pH 7.4
- Orbital shaker or tube rotator

Method

Carry out the following procedures in glass vials, unless otherwise indicated.

1. Wash free-floating sections twice in DEPC.PBS for 10 min each at room temperature with gentle agitation on a rotator or orbital shaker.

2. Acetylate sections in a solution of 25 mM acetic anhydride in 0.1 M triethanolamine in DEPC.PBS for 10 min with gentle agitation. Add the acetic anhydride only a few seconds before the sections.

3. Rinse sections for 5 min in DEPC.PBS.

4. Partially dehydrate sections by incubating in 70% ethanol (made up with absolute ethanol and DEPC.H$_2$O) for 2 min.

5. Pre-hybridize sections for at least 30 min at 37°C in hybridization buffer (*Protocol 10*, approx. 1 ml) without probe in an Eppendorf tube.

6. Continue with hybridization protocol for fluorescein labelled probes (*Protocol 14*).

5.2.3 *In situ* hybridization with fluorescein labelled oligonucleotides

We utilize an immunocytochemical method with an alkaline phosphatase-linked chromogenic reaction to detect the fluorescein labelled oligonucleotide after hybridization. This enzymatic reaction produces a blue-purple precipitate at the site of hybridization. Levamisole is an inhibitor of endogenous alkaline phosphatase which is added to reduce background.

Protocol 14. Hybridization, washing, and detection for fluorescein labelled oligonucleotide probes

Equipment and reagents

- Humidified chamber (for slide mounted sections): any closed assembly that allows slides to be laid absolutely horizontally with wet paper towels providing humidity will suffice (slide boxes from BDH, 406/0286/00 are suitable)
- Water-bath or oven set at 37°C for hybridization
- Water-bath set at 50°C for washing
- PAP pen (Dako, S2002) (for slide mounted sections)
- Glass vials and sterile Eppendorf tubes (for free-floating sections)
- Artist's paintbrush (for free-floating sections)
- Alkaline phosphatase-conjugated antibody to fluorescein (Amersham, RPN 3311)
- Levamisole (Sigma, L9756)
- Subbed microscope slides
- 0.2 × SSC (1/100 dilution of 20 × SSC with ultrapure water) (see *Table 1*)
- 1 × SSC (1/20 dilution of 20 × SSC with ultrapure water) (see *Table 1*)
- TBS: use stock solutions (see *Table 1*) to make 1 litre 100 mM Tris–HCl, 400 mM NaCl—adjust pH to 7.4, autoclave, and store at room temperature

- Hybridization buffer (*Protocol 10*)
- Blocking solution: dissolve 0.5 g Amersham Blocking reagent (RPN 3023) in 100 ml TBS—store in aliquots at –20°C
- Antibody buffer: 0.5% BSA (Sigma Fraction V, A3059) in TBS
- Detection buffer: 100 mM Tris–HCl, 100 mM NaCl, 50 mM $MgCl_2$ pH 9.5—check pH before using with high quality pH paper (Fluka Pehanon, 57008), and adjust with 1 M NaOH if necessary
- Stop buffer: 10 mM Tris–HCl pH 7.5, 10 mM EDTA containing 0.9% NaCl
- Glycerol gelatin: add 2 g gelatin (BDH) to 12 ml ultrapure H_2O. Warm gently on a heated stir plate in order to dissolve the gelatin. Add 14 ml glycerol (BDH) and 0.1 g thymol preservative. Warm with stirring for roughly 15 min until the flakes produced by the thymol have disappeared. To mount slides with this medium, heat it at 60°C for approx. 30 min until it is a runny solution. With cooling it becomes more viscous
- NBT (Boehringer Mannheim, 1 383 213): 75 mg/ml in 70% formamide
- BCIP (Boehringer Mannheim, 1 383 221): 50 mg/ml in dimethylformamide

A. *Hybridization*

1. Dilute fluorescein labelled probe into hybridization buffer to give final concentration of 0.5–1 pmol/ml.

2. Slide mounted sections. Set-up hybridization exactly as described in *Protocol 11*, hybridizing at 37°C overnight. Carry out post-hybridizaton washes with solutions in staining trays.

3. Free-floating sections. Transfer sections to an Eppendorf tube containing the probe in hybridization buffer (*Protocol 10*). Mix thoroughly and hybridize overnight at 37°C. Carry out the post-hybridization washes with solutions in glass vials.

B. *Post-hybridization washes*

1. Wash the sections for 2 × 15 min in 1 × SSC at room temperature.

2. Wash the sections for 2 × 15 min in 1 × SSC at 50°C for 15 min.

3. Wash the sections in 0.2 × SSC at 50°C for 15 min.

4. Wash the sections in 1 × SSC at room temperature for 30 min.

C. *Detection*

For slide mounted sections, put the slides into a humidified chamber. Use a PAP pen to make a hydrophobic boundary around the sections on the slide, and carry out the following incubations using about 500–800 μl buffer for each slide. For free-floating sections, carry out the following steps in Eppendorf tubes, or glass vials.

1. Wash the sections in TBS for 5 min.

2. Block non-specific binding sites on the sections with blocking solution for 30 min.

3. Dilute anti-fluorescein antibody coupled to alkaline phosphatase 1:1000 in antibody buffer and incubate the sections in this solution overnight at 4°C.

4. Wash the sections three times (10 min each) in TBS.

5. Wash slides in detection buffer for 5 min.

6. Prepare chromogen solution by slowly adding 45 μl NBT then 35 μl BCIP to 10 ml detection buffer. Finally add 2.4 mg levamisole. If using slide mounted sections, use about 500 μl chromogen solution for each slide, and incubate in the dark for 4–48 h.

7. Using a light microscope, monitor development at regular intervals. When the staining has reached the desired intensity, halt the reaction by incubation in stop buffer. Wash extensively in this buffer with frequent changes for 3 h to overnight.

8. For free-floating sections, mount them onto subbed slides. Allow to dry and coverslip using an aqueous mounting medium such as glycerol gelatin.

5.3 Combined *in situ* hybridization with immunocytochemistry

Using material prepared from perfusion fixed tissue, it is possible to combine *in situ* hybridization with immunocytochemistry. The quality of double-labelling depends crucially upon one's ability to obtain high quality single-labelled preparations using both techniques. Before attempting a double-labelling experiment therefore, it is advisable to master both immunocytochemical and *in situ* hybridization protocols independently. Once you have obtained good quality immunostaining, further dilution trials should be conducted to determine the concentration which gives the best signal-to-noise ratio for immunofluorescence under the conditions necessary to maintain

RNA integrity. We have found that the RNase inhibitor protein, RNasin, acts effectively to inhibit mRNA degradation during prior immunocytochemical incubations. RNasin requires dithiothreitol (DTT) to maintain its activity and we have noted that DTT can reduce labelling intensity with some antibodies. Thus, higher antibody concentrations may be necessary when combining immunofluorescence with *in situ* hybridization. Dilution trials for the immunocytochemistry should therefore include 0.5–1 mM DTT in antibody dilution buffers. We have found that RNasin at the concentrations we recommend does not have a detrimental effect on immunostaining, and inclusion of this enzyme in antibody dilution trials is therefore unnecessary (and expensive).

A method for combining radioactive *in situ* hybridization with fluorescent immunocytochemistry is given in *Protocol 15*. We find that protocols that do not end with fluorescent detection of antibodies (e.g. HRP; ABC) are not suitable for this since they cause chemography at the liquid emulsion stage (for a discussion of these artefacts see ref. 15). Don't forget to maintain RNase-free conditions from the very start of the experiment and throughout the immunocytochemical procedure (see Section 7 for details).

Protocol 15. Indirect immunofluorescence histochemistry combined with radioactive *in situ* hybridization

Equipment and reagents

- Humidified chamber: any closed assembly that allows slides to be laid absolutely horizontally with wet paper towels providing humidity will suffice (e.g. slide boxes from BDH, 406/0286/00)
- PAP pen (DAKO, S2002)
- Plastic staining dishes (BDH, 406/0236/00) and slide racks (BDH, 406/0237/00) treated with 1 M NaOH for 1 h then rinsed with DEPC.H$_2$O

- Primary antibodies directed against antigens of interest
- Secondary antibodies conjugated to fluorochrome (e.g. goat anti-rabbit IgG conjugated to fluorescein isothiocyanate; suppliers include Jackson ImmunoResearch, Sigma, and Molecular Probes)
- DEPC.PBS, 1 M DTT (see *Table 1*)
- Antibody dilution buffer: DEPC.PBS, 0.2% Triton X-100, 0.1% sodium azide

Method

1. As described in *Protocol 9*, cut thin sections (6–15 μm) of perfusion fixed frozen tissue using a cryostat and thaw-mount onto slides. Alternatively wash previously cut sections which have been stored in sucrose/ethylene glycol cryoprotectant five times for 10 min each in DEPC.PBS.

2. Using a PAP pen draw around the sections to form a hydrophobic barrier that will keep antibody solutions on the sections. Place the slides horizontally in the humidified chamber.

3. Dilute primary antibody to appropriate concentration using dilution buffer. You will need approx. 300 μl of antibody solution per slide.

4. Just before application of the primary antibody to the sections, add 1 M DTT to give a final concentration between 0.5–1 mM and then RNasin (Promega, N2511) at 100 U/ml.

5. Make sure all the sections are covered with antibody. Sections should never be allowed to dry or a high background or little specific staining will result. Place the lid on the chamber and tape it shut if necessary to ensure evaporation does not occur. Incubate the sections with primary antibody for 24–72 h at room temperature.

6. Keeping slides from drying, transfer to a slide rack and wash three times for 10 min each in DEPC.PBS (approx. 200 ml in a staining dish).

7. To appropriately diluted fluorescently-conjugated secondary antibody in dilution buffer add 1 M DTT (final concentration 0.5–1 mM) and RNasin (100 U/ml). Transfer the slides one by one to the humidified chamber and apply antibody solution (300 μl per slide). Seal chamber and allow to incubate 4–5 h at room temperature.

8. Transfer slides to racks and wash three times (10 min each time) in DEPC.PBS (approx. 200 ml in a staining dish).

9. Continue with pre-hybridization steps for radioactive *in situ* hybridization (*Protocol 8*C).

10. After pre-hybridization, take the slides directly to the hybridization, washing, and emulsion autoradiography steps (*Protocols 11, 12,* and *17*).

11. After development of the emulsion (*Protocol 18*), sections should be coverslipped using an aqueous mounting medium such as PBS: glycerol (1:3) containing 2.5% 1,4-diazabicyclo-[2,2,2]octane (DABCO, antifading agent) (Sigma, D2522).

5.4 Controls for *in situ* hybridization

Appropriate controls are essential to carry out high quality *in situ* hybridization (see also Chapter 1). In particular, it is essential to distinguish cells that are genuinely labelled from false positives (non-specific hybridization) and cells which are genuinely unlabelled from false negatives. Many controls have been used and can be found in the literature. One common control is the use of 'sense' probes, oligonucleotides which have a sequence which is the reverse complement of the probe of interest. These probes should give no signal in the tissue. We have found these of limited value since they do not directly determine whether the probe of interest shows specific binding.

Some controls are in-built due to the probe design. Probes which are specific for the mRNA of interest should be used. Even though it is always possible that a probe will bind to a related gene, as the genome databases grow, this prospect becomes increasingly unlikely.

For positive controls, we recommend the use of additional probes for comparative hybridization studies. It is easy to manufacture additional probes to compare the expression pattern of the target gene with a gene which is known to be expressed in the tissue of interest. In addition, synthesis of one or more additional probes to the same target will allow direct comparison of hybridization signal in the same tissues—if the two probes do not give identical patterns, there may be a problem. We do not recommend the use of control probes which hybridize to mRNAs that are expressed in all cells (e.g. actin, tubulin, GAPDH) since it is difficult to distinguish the pattern of hybridization using these probes with non-specific binding. If possible, therefore, use control probes that will specifically highlight a subset of cell types in the tissue. Another possible positive control for a probe is Northern blotting using a ^{32}P-labelled oligonucleotide probe (1). This method should show that an oligonucleotide binds to one or a small number of related mRNA species of a known length. However, since *in situ* hybridization techniques are very much more sensitive than Northern blotting, the lack of a specific band is not necessarily indicative of a poor probe, especially if the gene of interest is expressed at relatively low levels.

One excellent negative control that we use in every experiment is RNase pre-treatment of tissue sections before hybridization (*Protocol 16*). We find that, providing the probes are appropriately labelled, the biggest source of non-specific binding is from the tissue. This might be caused by inappropriate fixing or pre-treatment, or by the slides drying out, especially during the hybridization step. RNase pre-treated slides will usually reveal the extent of the problem—there should be no signal at all to RNase pre-treated slides. Another control is to dilute the labelled probe with an 50–100-fold excess of unlabelled probe—this control shows that the oligonucleotide will bind to sites which can be saturated, i.e. are present at low levels and are therefore likely to be specific. Similarly, experiments can be performed by increasing the temperature of the post-hybridization washes and then determining the pattern of hybridization. This last control is probably only important if the oligonucleotide probe has only a few mismatches with other genes.

Protocol 16.   RNase treatment of sections as a control

Equipment and reagents

- Two plastic or glass staining dishes reserved especially for this purpose
- Plastic slide racks reserved especially for this purpose
- Water-bath set at 37 °C
- Thermometer
- 50%, 70%, 95% ethanols

- 5 × RNase buffer: 2.5 M NaCl, 5 mM EDTA, 50 mM Tris pH 8; autoclave before use
- RNase A (Sigma, R-5503): make up about 10 ml of a solution of 10 mg/ml in 10 mM Tris pH 7.5, 15 mM NaCl and boil for 15 min—cool to room temperature and store frozen at –20 °C in 0.5 ml aliquots

Method

1. Make up RNase buffer with distilled water (100 ml 5 × RNase buffer plus 400 ml water) and place in staining dishes in water-bath. Into the first dish add RNase A to give a final concentration of 20 μg/ml. Allow solutions to reach 37 °C before proceeding (check with thermometer).

2. Remove slides from the freezer, and incubate slides in the dish containing RNase A at 37 °C for 30 min.

3. Incubate slides for 30 min at 37 °C in RNase buffer alone.

4. Dehydrate for 5 min in each of 50% ethanol, 70% ethanol, 95% ethanol.

5. Cover and air dry (10–60 min), and use for hybridization. The slides can be frozen at –70 °C until use.

6. Detection of hybrids and quantification

6.1 Detection of radiolabelled probes

6.1.1 Film autoradiography for radiolabelled probes

For low-resolution work, film autoradiography may be adequate. Slides should be loaded into autoradiography cassettes and film applied. X-ray film can be used for film autoradiography, but it is better to use the more sensitive films specifically designed for recording low energy beta emissions (e.g. Amersham βmax). For most oligonucleotides, an exposure time of one week is usually adequate to detect a signal, but for low abundance mRNAs or for quantitative work, longer times can be used to enhance signal strength. Films should be processed according to the manufacturer's instructions, preferably in vertical developing tanks, since it is easy to scratch the surface of the films when using horizontal tanks.

Unless the sections of tissue of interest are tiny, for a gene that is not expressed by all cells in the tissue, one will normally see a signal that is non-uniform over the sections (e.g. speckled), or at least is greater than that found in the RNase treated sections. By including ^{14}C-autoradiographic standards (Amersham), it is possible to carry out semi-quantitative analysis of films.

6.1.2 Liquid emulsion autoradiography for radiolabelled probes

For resolution of gene expression at the cellular level with radiolabelled probes, it is necessary to use liquid emulsion autoradiographic techniques which deposit a layer of emulsion a few micrometres thick directly on the tissue sections (*Protocol 17*). Radiolabelled probes convert silver halide salts in the photographic emulsion to small silver grains which can be viewed under the microscope to give cellular level resolution. We recommend the un-surpassed book by the late Dr A. W. Rogers (16). Whilst liquid emulsion autoradiography is not difficult to carry out, it is prone to failure since liquid emulsion can deteriorate and various parts of the procedure can give rise to artefacts.

Appropriate controls *are essential* to check that each stage of the procedure has worked. The first step is to carry out film autoradiography. If there is a high background over the RNase treated sections or slides, it is probably not worth carrying out liquid emulsion autoradiography. It is useful to emulsion dip blank slides along with the experimental slides which, when developed, will give an indication of emulsion background—if this is high then the most likely explanations are that the emulsion has gone off, or that there is light entering the dark-room where the procedures are carried out.

After exposure to emulsion, the slides are developed and may be counter-stained (*Protocol 18*). The optimal exposure time in liquid emulsion must be determined empirically, although for most of our work, four weeks is usually sufficient. Too short exposures will yield low (or no) signal, whereas over-exposure may produce saturation of the emulsion. By carrying out experiments with four or five extra slides, which are stored in a separate box, it is possible to develop slides at different time points and find the best exposure time for that experiment (the operating range). The conditions used for developing are critical to give a good signal with low background (16).

For quantitative experiments, it is essential that the entire experiment is dipped at the same time. The quality of the emulsion and developing conditions will vary from one experiment to the next, and therefore it is not possible to compare grain densities over cells from experiments that are not carried out and processed simultaneously.

Protocol 17. Liquid emulsion autoradiography

Equipment and reagents

- Dark-room
- Water-bath set to 42 °C (in the dark-room)
- Safelight (15 W bulb) with a filter suitable for the type of emulsion to be used[a] (for Ilford or Amersham emulsions an Ilford 904 filter is appopriate)
- Cupboard which can be sealed to light when people enter or exit the dark-room
- Racks or trays to allow microscope slides to be placed vertically whilst drying[c] (in the dark-room)
- Fume-cupboard
- Two 100 ml glass measuring cylinders (ultra-clean)
- Photographic emulsion[a] (e.g. Ilford K5), that has been stored away from radiation at 4 °C
- Dipping vessel (Amersham, RPN 39)
- Box of clean microscope slides with frosted ends
- Ultra-clean glass rod

- Plastic forceps
- Tightly sealed slide boxes for storage of slides (BDH, 406/0286/00)
- Aluminium foil
- Thermometer
- Gelatin capsules (Agar Scientific)
- Silica gel
- Acid for cleaning glass vessels. Either use chromic acid: 100 g potassium bichromate dissolved in 850 ml ultrapure water to which 100 ml concentrated sulfuric acid is added very slowly. This acid is extremely hazardous for skin, so gloves and face pro-tection should be worn when handling and should be handled and stored in fume-cup-board. Or use acid alcohol (less hazardous): 300 ml ethanol, 600 ml ultrapure water to which 100 ml concentrated hydrochloric acid is slowly added. Should be handled with gloves in fume-hood.

A. *Setting-up*

1. The day before, clean the measuring cylinders, glass rod, and dipping vessel. Wash for 2–3 h in 1 M NaOH and then in several changes of distilled water, scrubbing with a fine bottle brush to remove any flakes of old emulsion. In a fume-cupboard, the vessels can then be filled with chromic acid, or acid alcohol and left overnight. Then wash vessels with at least five changes of distilled water and check pH of liquid run-off to ensure that all of the acid has been removed. Dry vessels by inversion.

2. Switch the water-bath on in the dark-room.

3. Using a thermometer, check water-bath has reached correct temperature before proceeding. Double check that the dark-room is light-free, which may include taping over lights on the water-bath.

4. Fill one of the measuring cylinders with about 75 ml ultrapure water.

B. *Emulsion dipping*

1. First, place the cylinder containing water and the empty dipping vessel into the water-bath. Make sure that both vessels are secure and will not tip over into the water. It is best to put the dipping vessel in a small water-filled container within the water-bath so that any emulsion spilt can be easily cleaned up. It is not possible to leave the room from this stage on, so some attention to comfort will be necessary!

2. Switch off the light and under safelight conditions, open emulsion bottle.

3. Using forceps, remove threads of photographic emulsion and transfer them to the empty measuring cylinder. Typically, for a 50 ml bottle of Ilford K5 emulsion, remove about one-third of the bottle in order to dip 30–40 slides.

4. Place cylinder containing emulsion in water-bath for about 10 min. The emulsion will melt, and it will be possible to estimate the volume of the melted emulsion from the graduations on the measuring cylinder.

5. Add warmed ultrapure water to the cylinder containing the emulsion to dilute it 1:1.

6. Stir emulsion gently with the glass rod and leave for a further 5 min, stirring occasionally to help remove air bubbles.

7. Pour diluted emulsion into dipping vessel. Leave for a further 5 min.

8. To check that emulsion does not contain air bubbles, dip a clean plain-glass slide. Hold slide at the frosted end with the thumb and forefinger and, with an even movement, insert slide vertically into the dipping vessel. Slowly remove the slide and examine under the safelight. If the emulsion appears streaky or contains air bubbles, wait for a couple of minutes and test another slide.

Protocol 17. *Continued*

9. When emulsion is ready, dip the experimental sections. Dip each slide as described above, but count to three when the slide is fully inserted in the vessel. It is important that each slide receives an even layer of emulsion of similar thickness. This is a function of the time taken to dip the slide. Try to ensure that each slide is dipped in a similar fashion. If necessary, top up the emulsion in the dipping vessel, but each time remove air bubbles with plain slides. Dip some plain slides as controls.

10. Wipe back of slide with tissue to remove excess emulsion and place slides so that the emulsion can gel.[b]

11. Place slides in racks[c] in a light-proof cupboard and leave overnight to dry.[d]

12. Switch off water-bath, collect vessels etc., and leave room. Store unused emulsion at 4°C (used emulsion should not be reused: it will give a high background). Clean glassware, rod, and forceps thoroughly, before the traces of emulsion can dry.

13. The following day, place slides in slide boxes and wrap well with aluminium foil to prevent light entering. Carry out this step in complete darkness because the dry emulsion is more sensitive than liquid emulsion to stray light. Gelatin capsules containing silica gel desiccant can be included to reduce humidity.[e]

14. Leave slide boxes in a fridge (4°C) away from radiation for two to eight weeks (depending on signal strength) before developing (*Protocol 17*). If unsure about the exposure time, extra slides should be dipped that can be developed at different times to determine optimal exposure.

---

[a] In Europe, it is probably better to use Ilford K5 emulsion, or the similar Amersham emulsion. In North America, Kodak NTB-2 emulsion is more readily available. It should be noted that the conditions for each emulsion varies considerably, including the safelight filter, and the humidity and temperature for drying and exposing slides (16). The protocols here are not optimized for Kodak emulsions (see Chapter 3 for protocol).

[b] See ref. 16 for an explanation. How to best accomplish this depends on exact conditions available (especially dark-room space). Slides should be cooled to about 20°C so that the liquid emulsion gels before drying. This can simply be accomplished by hanging the slides vertically on a rack. Alternatively, cool the slides by allowing surplus emulsion to drip from the slides for a few seconds to achieve an even layer of emulsion and then placing them horizontally on a pre-cooled metal plate. Control experiments are necessary to find out the best conditions.

[c] It is easy to make a suitable rack by stringing clothes pegs on a small frame so that slides will hang completely vertically from their frosted ends without touching each other. Alternatively, we have used a commercial Perspex slide drying rack with success (BDH, 406/0273/00).

[d] The temperature and humidity at which the slides dry is important, but often difficult to control. For Ilford emulsions, this method works well if the emulsion dries at a relatively slow rate with moderate–high relative humidity (40–50%). If working without climate control, try to avoid dipping on very hot days. A Petri dish containing desiccant can be placed in the light-tight cupboard to help reduce humidity. Trial and error may be necessary to work out the best conditions to achieve consistency.

[e] The humidity and temperature of exposure is important. Rogers (16) recommends that desiccant is not used for Ilford emulsions, but one of us (G. J. M.) routinely uses desiccant with success.

Protocol 18.   Developing liquid emulsion slides

Equipment and reagents

- Dark-room equipped with a suitable safe-light
- Plastic staining racks, reserved for this purpose (BDH, 406/0236/00)
- Plastic slide racks (BDH, 406/0237/00)
- Thermometer
- Glass coverslips
- Photographic developing solution suitable for liquid emulsion (Kodak D19)
- Sodium thiosulfate

- Ultrapure water (500 ml for each batch of 25 slides)
- Toluidine blue stain: 0.2% toluidine blue dissolved in PBS—store in a glass, stoppered bottle and stir well before use
- Graded series of ethanols and xylene in glass staining dishes reserved for histological procedures
- DPX mounting medium

A. *Setting-up*

1. Make up developing solution according to the manufacturer's instructions. Do not use any metal with the emulsion powder (e.g. metal spatulas). The developing powder will dissolve slowly, and will have to be stirred for several hours to dissolve completely. Heating speeds up the powder dissolving but can cause problems. Developer can be made fresh, or better made up the day before. Undiluted developer can be stored in a dark vessel, but do not use if not clear, or if brown.

2. Make up a 25% solution of sodium thiosulfate. Stir until dissolved. 500 ml is needed for each batch of 25 slides.

3. Dilute developer 1:1 with ultrapure water. 250 ml is needed for each batch of 25 slides.

4. Heat (in water-bath) or cool (on ice) the solutions until they are exactly at 20°C.

5. Prepare five staining dishes, one containing 250 ml diluted developer, one containing 250 ml ultrapure water (wash 1), two containing 250 ml sodium thiosulfate each (fix 1 and fix 2), and one containing water (wash 2). Solutions should be used once only for each rack of slides.

B. *Developing sections*

1. In the dark-room under safelight conditions, unwrap slide boxes and place slides in a clean plastic slide rack.

2. Place rack in developer for exactly 3 min.

3. Place rack in wash 1 for 1 min.

4. Place rack in fix 1 for 6 min.

5. Place rack in fix 2 for 6 min.

6. Place rack in wash 2. Slides can now be exposed to light, but should be left in water.

7. For additional racks, replace the solutions and proceed as above.

Protocol 18. *Continued*

8. Wash slide racks in at least three changes of distilled water to remove excess thiosulfate.

C. *Counterstaining sections*

Unless slides have been processed for immunocytochemistry, it is usually convenient to counterstain tissue. Many stains are incompatible with liquid emulsion techniques because they are acidic and therefore cause negative chemography (removal of silver grains from the emulsion). We therefore suggest a simple Nissl stain (toluidine blue) dissolved in PBS (see refs 16 and 17 for details of some alternative stains).

1. Place slide racks in toluidine blue stain for 2.5 min.[a]

2. Place slides in distilled water for 5 min.

3. Dehydrate sections through a graded series of ethanols[b] and xylene[c] in glass staining dishes reserved for this purpose. We use 15 min in 50% ethanol (three times), 15 min in 70% ethanol (twice), 15 min in 95% ethanol (twice), 15 min in 100% ethanol (twice), and 15 min in xylene (twice, in fume-cupboard). It is important that the slides should not be removed from the ethanols or xylene and left in the air since they will become rehydrated.

4. Using a glass rod, spread a few drops of a permanent mounting medium (e.g. DPX, BDH 36029) onto a clean glass coverslip. Remove one slide from the xylene and place onto the coverslip. Press lightly, but firmly so that the mounting medium spreads over the whole surface of the slide. Ensure that no air bubbles are introduced.

5. Leave slides overnight for the mounting medium to harden. The next day remove all the emulsion from bottom surface of slide by spraying with distilled water or 70% ethanol and scraping carefully with a single-edged razor blade. Remove excess dried mounting medium with the razor blade.

6. Spray slides with 70% ethanol and clean with tissue before microscopy.

---

[a] Stains which give a very strong colour can affect the viewing of sections under dark-field microscopy (although this problem is largely overcome through use of a polarization filter on the microscope). If dependent on dark-field microscopy, it is likely to be necessary to sometimes adjust the strength of the stain or the staining time to achieve optimal results. With some experience, it is usually possible to judge the quality of the staining by the time the slides reach the first 70% ethanol stage. Understained sections can be rehydrated by moving them back through 50% ethanol, water, and can then be restained for a longer time. Overstained sections can also be rehydrated and left in water for 30 min–1 h, this will sometimes remove enough stain to give better viewing with dark-field.

[b] The ethanols can be reused a number of times if the staining dishes are sealed with Parafilm, but should be replaced when they become very discoloured.

[c] Xylene should be stored in stoppered glass bottles when not in use. Water can be removed from xylene by filtration through a phase separation filter (e.g. Whatman grade 1PS). Caution: xylene is harmful by inhalation.

6.2 Imaging and analysis of sections hybridized with radiolabelled probes

6.2.1 Microscopy of liquid emulsion autoradiograms

Slides that have been used for liquid emulsion autoradiography should be viewed with a light microscope. Using a high magnification objective ($\times 25$ or above) it will be possible to see silver grains (small dark spots) and stained cells (blue profiles). For many cell types, a positive cell will be indicated by clusters of grains above the blue stained Nissl substance that fills the cytoplasm. However, some cell types are not well stained by toluidine blue under these conditions, and it may be necessary to try a variety of stains (16, 17).

Grains can also be visualized at lower magnification by using a dark-field condenser on the microscope. Here, silver grains are seen as bright white spots and the underlying cells should not be visible. Using dark-field microscopy the underlying tissue, particularly if heavily counterstained, will sometimes show, or interfere with the brightness of the grains. This problem may complicate analyses since silver grains may be difficult to resolve from tissue background signals. This problem can be overcome somewhat by using more lightly counterstained material. However, to eliminate this problem entirely slides can be viewed using a microscope with a high intensity light source (as used for fluorescence microscopy) where the light source is passed through a polarization filter block (18). Here, the grains appear white or light blue and the background cells can be viewed simultaneously.

6.2.2 Semi-quantitative studies using liquid emulsion autoradiographs

In situ hybridization can be used as a semi-quantitative technique to measure the level of gene expression in cell types or tissues. Under the right conditions, the number of silver grains present is related to the amount of radioactivity that has been bound to a cell (16), and the amount of radioactivity is related to the amount of mRNA of interest in that cell. This relationship has been shown to be true for a few probes (2, 19), but can be assumed to be true for all oligonucleotide probes since they are a single molecular species which will bind reversibly to a single target. Quantification of *in situ* hybridization, therefore, is analogous to measurement of receptor number using radioligand binding techniques (20). In practice, to estimate the number of mRNA species per cell is possible, but not practicable (19), since it will involve carrying out hybridization experiments with several concentrations of probe. It is, however, relatively easy to measure differences in the cellular level of expression of mRNAs, for example to compare the effect of a drug or surgical treatment, or to determine whether the level of gene expression in one cell population is different from another. This is termed semi-quantitative analysis, since the level of mRNA determined is not an exact number of molecules per cell, but is relative to a parameter internal to the experiment. Ideally, this type of experiment should be carried out using a concentration of oligonucleotide probe that is saturating so that all of the available binding sites (mRNA targets) are bound to probe, and the number of silver grains is therefore directly related to the amount of mRNA present per cell. In practice, the concentration of probe required to achieve saturation will make this approach impractical due to expense, and there will be high non-specific binding under these conditions. It is therefore usual to use a lower concentration of probe that, ideally, would represent a concentration of probe around the K_d (analogous to receptor autoradiography on tissue sections) (21). Under these conditions, the number of silver grains per cell is related to the amount of mRNA present, with a near-linear relationship. For the few probes that we have measured, a concentration of about 2 nM is close to the K_d (2, and unpublished observations).

The relationship between the amount of radioactivity present and the number of silver grains is dependent on the operating range of the photographic emulsion used. If there is 'saturation' of the emulsion (overexposure), there is no longer a linear relationship between the number of grains and the amount of mRNA in a cell. When carrying out semi-quantitative *in situ* hybridization it is important to define the operating range, which can be done empirically by emulsion dipping extra sections to determine the optimum exposure time.

If using *in situ* hybridization for semi-quantitative studies, it is essential to design the experiments well to minimize variation. We always carry out hybridization, washing, emulsion dipping, and developing simultaneously for all slides to be used in one experiment, and include RNase controls and (usually) multiple slides from each tissue. For each experimental group, tissues from at least six independent cases should be taken, to allow meaningful statistical tests to be done. Often this will mean carrying out an experiment on 60–72 slides at one time.

6.2.3 Image analysis liquid emulsion autoradiographs

The simplest form of image analysis is by eye (16). Using a microscope with a high magnification objective ($\times$ 25 or $\times$ 40), equipped with a graticule (it is best to use a grid that fits in the eyepiece), it is possible to count the grains over cells and over equivalent sized areas of background taken from between cells. For cells that are scattered throughout the section, using ^{35}S-labelled probes and Ilford K5 emulsion, we recommend a box size that is 10–20 μm greater in diameter than the average diameter of the cell, to allow for scatter of grains from a point source (16). For cells which are in layers or close together, it may be more useful to use boxes that are smaller than the mean diameter of the cell (16). To estimate background, at least five background fields should be taken from each image between the cells. Using this figure, each cell can be defined as positive or negative depending on the number of grains over a cell, related to the number of grains over a similar area of background. Some authors define an arbitrary cut-off, say a grain density twice, five or ten times the background grain density. We prefer to define a positive cell by taking the mean of the background density plus twice the standard deviation of the background density. If a cell has a grain density above this value, then we have >95% confidence that this cell is genuinely positive.

It is possible to speed up the analysis by using a computer-aided system. For this the microscope must be fitted with a video camera that is linked to a computer equipped with suitable hardware. An image is 'captured', i.e. converted into a digital form that can be analysed by the computer. The image is broken down into a grid of pixels, usually 512 × 512, each pixel having a value according to its greyness—for a black and white camera this value this is between 0–255. Analysis software then controls the extraction of information from that image. There are, essentially, two different ways of measuring grain density using computer-aided image analysis. The first type of method (e.g. ref. 22) uses a bright-field image at high magnification (× 40 or × 25). The silver grains are distinguished from the underlying cells because they are darker (the pixels which make up the grains have a lower grey value). This process is called 'thresholding'. Typically, using a mouse, a box of appropriate size (see above) is placed over a cell and the computer then calculates the area of that box that has been thresholded (i.e. is occupied by grains). Some programs will then use this value to calculate the exact number of grains. In the second method, a bright-field image can be captured and used to place boxes over cells. Without moving the slide on the microscope, the illumination is then changed to dark-field or epipolarization. Here the grains are lighter than the overlying image and can be thresholded on that basis. The computer calculates the area of the boxes overlying the cells that is occupied by silver grains. For this method, it is possible to carry out the analysis at lower magnification (× 16) so that more cells can be analysed from each image captured.

For either method, the data can be exported to a spreadsheet program that can be used to calculate:

(a) The grain density of the background.
(b) Whether a cell is positive or not (grain density greater than background mean + 2 × standard deviation).
(c) The mean grain density of positive cells.
(d) The proportion of cells analysed which are positive.

A number of computer programs exist that can be modified to carry out one or both of these methods, but a full description of them is beyond the scope of this chapter. We have customized one such system, '*Visilog*' (Noesis), that provides a sophisticated analysis platform at low cost, but more sophisticated, easier to use (and generally more expensive) systems are available.

6.2.4 Film autoradiography

For radioactive probes, film autoradiography will often give an image of the required resolution when localizing the expression of a mRNA to a particular tissue. In this case, RNase controls are especially important to determine the degree of non-specific binding of the probes for that particular experiment. Image analysis can be carried out, provided autoradiographic standards have been used, and the experiments have been carried out within the 'operating range' of the film, i.e. that the film is not saturated (21).

6.3 Image analysis using non-radioactive probes

The results of non-radioactive *in situ* hybridization experiments are viewed through a light microscope using bright-field optics. Using image analysis software such as that described above, it is also possible to measure RNA levels in cells hybridized with fluorescein labelled oligonucleotides. As for all other *in situ* methods, there are a number of provisos that are necessary to ensure good semi-quantitative methods. In this case, it is important to ensure that the sections are incubated in substrate under conditions that produce a near-linear relationship between starting material and product. Of special importance is that the light source should be stable and reproducible. It may be necessary to quantify a number of cells from a standardized slide in each counting session to correct for subtle differences in illumination. Methods for quantification of alkaline phosphatase detection of oligonucleotide binding have been described (23).

7. Stock solutions and precautions against RNase contamination

Unless precautions are taken, degradation by RNase will prevent successful *in situ* hybridization. It is important to prepare solutions and carry out laboratory procedures in a way that minimizes the possibility of RNase contamination. Stock solutions should be made up by weighing reagents into tissue culture grade sterile plasticware or oven baked (250 °C overnight) glassware. Magnetic stir bars and spatulas should be autoclaved, and water used for solutions should be ultrapure and sterile wherever possible. Many aqueous solutions can be treated to remove RNase with diethylpyrocarbonate (DEPC, BDH 44170), with the exception of Tris solutions. After preparation, we recommend that when possible aqueous solutions should be sterilized in an autoclave before use. This will prolong their shelf-life. We recommend that solutions are stored in aliquots, and gloves must be worn when handling. Similarly, sterile pipette tips and Eppendorf tubes should be used. It is essential that plastic staining dishes are properly cleaned after each use, particu-

Table 1 Stock solutions for *in situ* hybridization with oligonucleotides

DEPC. A potential carcinogen, so must be handled in a fume-hood and appropriate protective clothing worn. To prepare DEPC treated H_2O, PBS, 0.9% NaCl, 20 × SSC, sodium acetate, or phosphate buffer, add 1 ml DEPC per litre of solution, shake well, leave overnight, and autoclave.

100 × Denhart's solution. To 5 g Ficoll (Type 400, Pharmacia), 5 g polyvinylpyrrolidone, and 5 g bovine serum albumin, add 200 ml DEPC.H_2O and stir overnight in a beaker that has been previously oven baked, using a stir bar that has been previously autoclaved. Make up to a final volume of 250 ml and store in 15 ml aliquots at –20 °C, where it is stable for several years.

1 M dithiothreitol. Make up 1.54 g dithiothreitol (Sigma, D9779) to a volume of 10 ml with DEPC.H_2O in an oven baked volumetric flask. Store at –20 °C in 0.5 ml aliquots in sterile Eppendorf tubes.

Denatured, sheared herring sperm DNA. Use high quality DNA (Boehringer Mannheim, 223 646). Using dissecting scissors cut about 200 mg of the DNA into sterile 50 ml conical tube. Weigh the DNA, and add ultrapure water to give a concentration of about 10 mg/ml. Heat and vortex occasionally until DNA is dissolved (may take 2–3 h). Pass the solution 12 times through syringe fitted with 17 gauge needle to shear the DNA. Denature the DNA by boiling the solution for 10 min and place immediately on ice to prevent renaturation. Using a spectrophotometer, calculate the DNA concentration and adjust the volume with ultrapure water if necessary to give a concentration of 10 mg/ml. Store in 2.5 ml aliquots at –20 °C. Stable for months–years.

0.5 M EDTA. Add 18.6 g disodium EDTA.$2H_2O$ to 80 ml ultrapure water. Stir and heat, and adjust to pH 8 by adding sodium hydroxide pellets. Continue stirring until the solution clears. Check pH, adjusting if necessary, make up to 100 ml, and autoclave.

Deionized formamide. To 500 ml high quality formamide (BDH, 10326), add 5 g of a mixed-bed resin (Dowex MR3, Sigma I-9005). Stir for 2 h in an oven baked glass beaker in a fume-cupboard. Allow resin to settle, decant into sterile 50 ml tubes, and store at –20 °C. Do not use formamide if it is yellowish in colour. Formamide is potentially a teratogen.

0.9% NaCl. Dissolve 9 g NaCl in 800 ml ultrapure water. Adjust volume to 1 litre. Autoclave before use.

4 M NaCl. Dissolve 233.8 g NaCl in 800 ml water. Adjust volume to 1 litre. Autoclave before use.

PBS. Dissolve 8 g NaCl, 0.2 g KCl, 1.4 g Na_2HPO_4, and 0.24 g KH_2PO_4 in 800 ml ultrapure water. Adjust to pH 7.4 with NaOH, and make up to 1 litre with ultrapure water. Autoclave before use.

0.2 M phosphate buffer pH 7.4. Prepare by combining approx. 19 ml 0.2 M monobasic sodium phosphate (31.2 g $NaH_2PO_4.2H_2O$ in 1000 ml H_2O) with approx. 81 ml 0.2 M dibasic sodium phosphate (71.7 g $Na_2HPO_4.12H_2O$ in 1000 ml H_2O) until a final pH of 7.4 is achieved. Autoclave before use.

3 M sodium acetate pH 5.2. Dissolve 40.8 g sodium acetate ($C_2H_5ONa.3H_2O$) in 70 ml water. Adjust to pH 5.2 with glacial acetic acid (at least 15 ml will be necessary). Adjust volume to 100 ml.

20 × SSC. Weigh out 175.3 g NaCl and 88.2 g sodium citrate. Add 800 ml ultrapure water and stir until dissolved. Adjust to pH 7 with sodium hydroxide and make up volume to 1 litre. Autoclave before use.

1 M Tris (various pH). Dissolve 121.1 g Tris base in 800 ml ultrapure water. Add concentrated HCl (70 ml for pH 7.4, 65 ml for pH 7.5, 60 ml for pH 7, 42 ml for pH 8, 0 ml for pH 9.5). Make up to 1 litre and autoclave. Make sure that the pH meter has an electrode that is suitable for Tris buffer. Otherwise, use high quality pH papers (Fluka). Do not treat Tris solutions with DEPC.

larly if they are not reserved for this technique. 1 M NaOH for at least 1 h followed by several rinses in DEPC.H$_2$O is effective in reducing the level of RNase.

References

1. Sambrook, J., Fritsch, E.F., and Maniatis, T. (ed.) (1989). *Molecular cloning: a laboratory manual*, 2nd edn. Cold Spring Harbor Laboratory Press, Cold Spring Harbor, NY.
2. Savery, D., Priestley, J.V., and Rattray, M. (1992). *Biochem. Soc. Trans.*, **20**, 304S.
3. Lewis, M.E., Sherman, T.G., and Watson, S.J. (1985). *Peptides*, **6 (Suppl. 2)**, 75.
4. Program Manual for the Wisconsin Package, Version 8, September 1994. Genetics Computer Group, 575 Science Drive, Madison, Wisconsin 53711, USA.
5. Matteucci, M.D. and Caruthers, M.H. (1981). *J. Am. Chem. Soc.*, **103**, 3185.
6. Altschul, S.F., Gish, W., Miller, W., Myers, E.W., and Lipman, D.J. (1990). *J. Mol. Biol.*, **215**, 266.
7. Kiyama, H. and Emson, P.C. (1990). *Neuroscience*, **38**, 223.
8. Schmitz, G.G., Walter, T., Seibl, R., and Kessler, C. (1991). *Anal. Biochem.*, **192**, 222.
9. Kawata, M., Yuri, K., and Kumamoto, K. (1990). *Acta Histochem. Cytochem.*, **23**, 307.
10. Dirks, R.W., Van Gijlswijk, R.P., Vooijs, M.A., Smit, A.B., Bogerd, J., van Minnen, J., *et al.* (1991). *Exp. Cell Res.*, **194**, 310.
11. Syrjanen, S., Henonen, O., Miettinen, R., Paljarvi, L., Syrjanen, K., and Riekkinen, P. (1991). *Neurosci. Lett.*, **130**, 89.
12. Asanuma, M., Ogawa, N., Mizukawa, K., Haba, K., and Mori, A. (1990). *Res. Commun. Chem. Pathol. Pharmacol.*, **70**, 183.
13. Dagerlund, A., Friberg, K., Bean, A.J., and Hökfelt, T. (1992). *Histochemistry*, **98**, 39.
14. Khan, G., Coates, P.J., Kangro, H.O., and Slavin, G. (1992). *J. Clin. Pathol.*, **45**, 616.
15. Priestley, J.V., Wotherspoon, G., Savery, D., Averill, S., and Rattray, M. (1993). *J. Neurosci. Methods*, **48**, 99.
16. Rogers, A.W. (1979). *Techniques of autoradiography*, 3rd edn. Elsevier, Amsterdam.
17. Schnell, S.A. and Wessendorf, M.W. (1995). *Histochem. Cell Biol.*, **103**, 111.
18. Priestley, J.V. (1992). *Microsc. Anal.*, **32**, 15.
19. Nunez, D.J., Davenport, A.P., Emson, P.C., and Brown, M.J. (1989). *Biochem. J.*, **263**, 121.
20. Bennett, J.P., Jr. and Yamamura, H.I. (1985). In *Neurotransmitter receptor binding* (ed. H.I. Yamamura, S.J. Enna, and M.J. Kuhar), 2nd edn, pp. 61–89. Raven Press, New York.
21. Kuhar, M.J. (1985). In *Neurotransmitter receptor binding* (ed. H.I. Yamamura, S.J. Enna, and M.J. Kuhar), 2nd edn, pp. 153–76. Raven Press, New York.

22. Masseroli, M., Bollea, A., Bendotti, C., and Forloni, G. (1993). *J. Neurosci. Methods*, **47**, 93.
23. Asan, E. and Kugler, P. (1995). *Histochemistry*, **103**, 463.

3

Detection of mRNA in tissue sections with radiolabelled riboprobes

ANTONIO SIMEONE

1. Introduction

Gene expression studies have been greatly advanced by techniques developed to visualize gene expression during embryonic development and tissue differentiation. Among these, *in situ* hybridization of sections with radiolabelled probes has a historical significance because the *in situ* expression of mRNA in invertebrate and vertebrate development was first described using this procedure. Although alternative techniques are now available to identify gene transcripts in tissues and embryos (1, 2), *in situ* hybridization on sections with radiolabelled RNA probes is frequently found to be the more sensitive and reliable procedure. This chapter describes protocols for *in situ* hybridization with radiolabelled RNA probes on embryo and brain sections. In addition, a list is provided of diagnostic probes useful at specific stages of development for defining the expression domains of a given gene.

2. Recovery and fixation of embryos

The protocols described below have been used extensively by us for the analysis of gene expression during mouse development, and some information is provided regarding the handling of these embryos. These methods are applicable to many embryonic systems and tissues with some simple adjustments, for example allowing for the size of the tissue during embedding for sectioning.

If many genes need to be analysed at different stages it is a good strategy to collect a large number of embryos from critical stages of development (3, 4) so that the experiments can be carried out quickly. Since mouse embryos grow extensively during development, in order to generate sufficient sections, more embryos should be collected from earlier stages than from later stages. A good working stock of mouse embryos is listed in *Table 1*. This stock of embryos assures the expression analysis of at least ten genes during gastrulation, and 20–25 during neural patterning and somitogenesis, as well as during later organogenesis.

Table 1. Working stock of mouse embryos

	Gastrulation		Neural patterning and somitogenesis		Organogenesis (differentiation of tissues)		
Stage (d)	6.5	7.5	8.5	10.5	12.5	15.5	17.5
Number of embryos	40–50	40–50	30–40	20–25	10–12	6	6

The procedure for the dissection and fixation of mouse embryos is given in *Protocol 1*. It should be noted that for embryos younger than 9.5 d (days post-coitum), it can be more convenient to fix them in the decidua since they are then easier to manipulate and dissect. After fixation, embryos are washed to eliminate residual PFA and partially dehydrated in ethanol where they can be stored indefinitely at 4 °C. 8.5–9.5 d mouse embryos can be dissected from the decidua when they are in 70% ethanol.

Protocol 1. Fixation and storage

Equipment and reagents
- Stereo dissecting microscope
- PBS: 8 g NaCl, 0.2 g KCl, 1.44 g Na$_2$HPO$_4$, 0.24 g KH$_2$PO$_4$ dissolved in 1 litre of distilled water and pH adjusted to 7.4—sterilize by autoclaving
- PFA: 4% paraformaldehyde dissolved in PBS at 55–65 °C and then cooled on ice
- Saline solution: 8.3 g of NaCl in 1 litre of distilled DEPC treated water—sterilize by autoclaving

Method
1. Dissect out embryos in a Petri dish under stereomicroscope in ice-cold PBS.
2. Transfer the embryos to a fresh Petri dish containing PBS and wash for a few minutes, shaking very slowly.
3. Transfer the embryos to a vial containing ice-cold freshly prepared PFA[a] and leave at 4 °C for 8–16 h.[b]
4. Replace PFA with saline solution and wash two times with gentle rotation for 30–90 min[c] at 4 °C.
5. Wash with 1:1 saline solution:ethanol for 30–90 min[c,d] at 4 °C.
6. Wash with 70% ethanol for 30–90 min[c] at 4 °C.
7. Repeat step 6 and store at 4 °C.[d]

[a] All the steps from fixation to wax embedding should be performed using at least 10 ml of solution for three to eight early embryos (6.5–10.5 d), 20 ml for two to four mid-gestation embryos (12.5–14.5 d), and 20–30 ml or more for one or two embryos of later stages.
[b] Younger embryos from 8.5–10.5 d without the decidua should be fixed for no more than 12–16 h; they are generally already fixed after 6–8 h. For older embryos (15–19 d), fixation should be prolonged to 48 h. In this case it is wise to replace fixative after 20–24 h. To improve fixation of very late embryos, cut carefully into the belly and the vault of the cranium.
[c] Washes are increased in length depending on the size of embryos.
[d] Embryos can be stored in 70% ethanol indefinitely at 4 °C.

---

3. Wax embedding, orientation, and sectioning

We use wax embedding (*Protocol 2*) for analysis of mouse embryos up to late stages of gestation. For very late stages, or adult tissues, cryostat sections are more suitable (see Section 8). Before being embedded in wax, embryos are completely dehydrated through increasing concentrations of ethanol in order to remove all traces of water which cannot be substituted by wax. 100% ethanol is in turn replaced by toluene which is finally replaced by melted paraffin wax. Dehydration and toluene treatment can be performed at 4 °C but identical results are obtained at room temperature. As for the fixation procedures, the various steps should be performed for different lengths of time depending on the size of embryos. For early stage mouse embryos (7–10.5 d) 30 min are sufficient, but for late stage embryos (16–18 d) 1–2 h are necessary for each step.

A particularly critical step before sectioning the embryos is the correct orientation of the embryos as this greatly facilitates the identification of anatomical areas. Moreover, it is advisable, at least in a preliminary analysis, to confirm hybridization in a given area or tissue by analysing several planes of section. Sagittal, transverse, and coronal views assure a reliable and defini-tive identification of the signal, but this exhaustive analysis is time-consuming if multiple stages and different plans are analysed separately. To overcome this problem we set up a general strategy to elucidate the gene expression pattern during post-implantation development by analysing different planes at the same time of the critical developmental stages: 7.5, 8.5, 10.5, 12.5, 15.5, and 17.5 d embryos.

For very early stages, eight 7.5 d and six 8.5 d embryos are embedded in the same mould by orienting them in equal number for transverse and sagittal sections. For these stages at least three to four embryos per plane are required. Orientation is critical, as depicted in *Figure 1*. Similarly, three 10.5 d and 12.5 d embryos, and two 15.5 d and 17.5 d embryos, are embedded together. At 10.5 d and 12.5 d, embryos are oriented in sagittal, coronal, and transverse views; for later stages, sagittal and coronal views are sufficient (*Figure 2*). Embryos are sectioned to produce ribbons of 7 μm per section. One by one, sections are carefully placed on slides (pre-treated as described in *Protocol 3*) to produce a series of slides of adjacent sections (*Protocol 4*). This will generate a relatively high number of slides for mid–late stage embryos and three to six slides for earlier stages (*Table 2*). One slide of each stage should be used for each probe and the total number of slides should correspond to about 50 in total. The main advantage of this strategy is that additional genes to be analysed can be hybridized on adjacent sections, facilitating combinatorial expression analysis. The effort expended in these systematic embedding and sectioning steps will be amply repaid by the easier and more exhaustive analysis with a major saving of time.

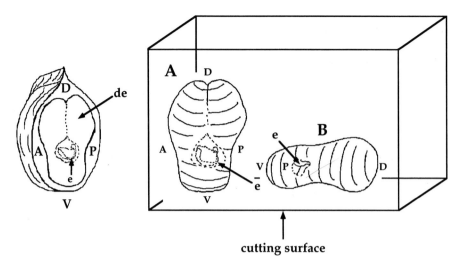

Figure 1. Schematic representation of a gastrulation stage mouse embryo (e) in the deciduum (de) as it appears in the uterus (on the left), and of two embryos in decidua orientated in a wax cube (on the right) to obtain transverse (A) and sagittal (B) planes in the same series of sections. A, P, D, V indicate anterior, posterior, dorsal, and ventral sides, respectively.

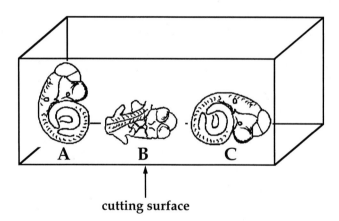

Figure 2. Schematic representation of three 10.5 d embryos orientated in a wax cube to obtain three different orientations in the same series of sections. Embryos A, B, and C will be sectioned in transverse, sagittal, and coronal planes, respectively.

Table 2. Number of slides per series at different embryonic stages

Stage (d)	6.5	7.5	8.5	10.5	12.5	15.5	17.5
Number of slides per series	2	3	3–4	5–6	8–10	11–13	14–18

Protocol 2. Wax embedding

Equipment and reagents
- Incubation oven at 60°C
- Heater block at 60°C
- Stereomicroscope
- Moulds for wax embedding
- Paraffin wax melted at 60°C

Method

1. Replace 70% ethanol with 85%, then 95%, and finally twice with 100% ethanol.[a] Each wash is at 4°C for 30–90 min with gentle rotation.

2. Replace ethanol with toluene at room temperature for 30–90 min with gentle rotation.

3. Repeat step 2.

4. Replace toluene with a 1:1 toluene:paraffin wax solution[b] and keep in an incubation oven at 60°C for 30–90 min with occasional agitation.

5. Replace with melted paraffin wax and keep at 60°C for 30–90 min with occasional agitation.

6. Repeat step 5 two or three times.

7. Transfer embryos to a mould together with enough paraffin wax to submerge the embryo(s).

8. Place mould on heater block at about 60°C, and under a stereomicroscope orientate the embryos using a warmed needle (see *Figures 1* and *2*). Cool to room temperature and store at 4°C.

[a] Embryos can be stored overnight in 100% ethanol; this is recommended only for late stage embryos (15.5–19.5 d).
[b] Paraffin wax is melted, filtered through filter paper inserted in a funnel in a bottle, and can be stored in a heater at 60^°C for up to a few days.

Protocol 3. Subbing of slides

Equipment and reagents
- Slide racks for 25 slides
- Oven
- TESPA (aminopropyltriethoxysilane) (Sigma, A3648)
- Crome-alum gelatin: 0.5% crome-alum, 0.5% gelatin. Dissolve gelatin at a concentration of 5 g per litre in DEPC treated water at 60°C on a stirrer, and then add 5 g chrome-alum per litre. Filter through Whatman 3MM paper.

Method A

1. Put slides in slide rack(s).

2. Dip for 30 sec in 10% HCl, 70% ethanol.

Protocol 3. *Continued*

3. Wash in distilled water.

4. Dip for 30–60 sec in 95% ethanol.

5. Dry in an oven at 70–80 °C for 20–40 min and allow to cool.

6. Dip slides in 2% TESPA in acetone for 30 sec.

7. Wash twice in acetone and then in distilled water.

8. Dry at 37–42 °C for several hours. Can store airtight for months at room temperature.

Method B

1. Wash slides with an abundant volume of 0.5% SDS.

2. Wash slides in tap-water for 1 h.

3. Put slides in 25 slide rack(s) and wash twice in bidistilled water.

4. Dip slides in chrome-alum gelatin for 2 min.

5. Air dry and bake at 150 °C for 2–3 h or at 37 °C overnight. Can store airtight for months at room temperature.

Protocol 4. Sectioning

Equipment

- Microtome
- Slide warmer

Method

1. Trim the wax cube to a truncated pyrimidal shape, such that the surface to be cut is reduced to an appropriate size.

2. Attach to the microtome, align relative to the blade, and set the section thickness to 7 μm.

3. Cut ribbons and place on filter paper, maintaining the order of sections.

4. Label the series of slides, place them on a slide warmer at 40°C, and put 2–3 ml 10% ethanol on each. The number of slides per series is determined on the basis of the embryonic stage (*Table 2*).

5. Remove a single section from the ribbon.

6. Float this section onto a slide.

7. Place the next section onto the next slide (steps 5 and 6) for the rest of this series. Then repeat until this first series of slides is filled with sections.

8. Put subsequent sections on a new series of slides until all the sections have been mounted.

4. Probe preparation

The correct preparation of the single-stranded RNA probe (*Protocol 5*) is critical for a good final result. For this reason particular attention should be given to this phase, which is now relatively easy using well established protocols and nucleotide mix solutions. For good probes a general recommendation is that the templates should be 0.3–1 kb long (see Chapter 1). The template sequence is cloned into a vector so that it is flanked by promoters for T3, T7, or SP6 RNA polymerase. To obtain an antisense RNA probe the plasmid must be linearized using a restriction enzyme that cuts 5′ of the insert, and then transcribed with the RNA polymerase promoter located at the 3′ of the insert. We recommend purifying the probe from unincorporated nucleotides by using a Sephadex quick spin column. The final yield of labelled probe should be sufficient to analyse 20–40 slides using a concentration of ~ 3–5×10^4 c.p.m./μl of hybridization mix (see below).

Protocol 5. Synthesis of single-stranded RNA probe

Reagents

- 10 × transcription buffer: 400 mM Tris–HCl pH 7.5, 60 mM MgCl$_2$, 20 mM spermidine, 100 mM NaCl
- Nucleotide mix: 2.5 mM ATP, 2.5 mM GTP, and 2.5 mM UTP
- TE: 50 mM Tris–HCl, 1 mM EDTA pH 8— sterilize by autoclaving
- Phenol:chloroform: mix phenol, chloroform, isoamyl alcohol in the ratio 25:24:1— store at 4°C

Method

1. Mix the following reagents at room temperature:

   - 10 × transcription buffer   2 μl
   - 100 mM dithiothreitol (DTT)   2 μl
   - recombinant RNasin ribonuclease inhibitor (20 U/μl)   1 μl
   - nucleotide mix   4 μl
   - 100 μM CTP   2 μl
   - linearized template DNA (1 μg/μl)   1.5 μl
   - [α-^{35}S]CTP (1000 Ci/mmol, 10 mCi/ml)   6 μl
   - RNA polymerase (20 U/μl)   2.5 μl

2. Incubate for 60–90 min at 37°C.

3. Add RNase-free DNase at 1 U/μg of template DNA.

4. Incubate 30–45 min at 37°C.

5. Add 300 μl of 1 M ammonium acetate and 20 μg of tRNA.

6. Add 2 vol. of phenol:chloroform, vortex, and spin for 2 min in a microcentrifuge.

Protocol 5. *Continued*

7. Recover the aqueous phase and precipitate RNA probe by adding 2.5 vol. of 100% ethanol, mix, and chill for 10 min at −70°C. Spin for 5 min in a microcentrifuge.

8. Redissolve the pellet in 100 μl TE.

9. Load on a 1 ml Sephadex G50 spun column, centrifuge 10 min at 2000 r.p.m., and recover the eluate (Eppendorf centrifuge 5810R).

10. To partially hydrolyse the probe add 0.1 vol. of freshly prepared 1 M sodium hydroxide and keep on ice for 10–20 min.[a]

11. Precipitate RNA by adding 20 μg tRNA, 0.1 vol. of 3 M sodium acetate pH 5.5, and 2.5 vol. of 100% ethanol, and chill for 10 min at −70°C. Spin in microcentrifuge for 5 min.

12. Redissolve the pellet in 100 μl of TE, mix 1 μl of probe in scintillation liquid, and determine the amount of radioactivity (c.p.m.).[b]

---

[a] To improve the penetration of probe into tissues, the length of RNA probe can be reduced to 0.1–0.25 kb (see Chapter 1). The easy procedure described here gives good results, but a different method is recommended for a more precise hydrolysis to the correct length (5) (see Chapter 4).
[b] A good probe yields between 8×10^5 and 1.5×10^6 c.p.m./μl.

5. Pre-treatment and hybridization

Before undergoing hybridization, sections are treated with xylene to completely remove the paraffin, then with proteinase K to improve penetration of the probe, and finally with acetic anhydride to reduce background due to electrostatic binding of probe (and also inactivate residual protease) (*Protocol 6*). The xylene and acetic anhydride treatments pose no problems, but the proteinase K treatment is a critical step to control since higher or lower activity of the enzyme can result in subsequent problems in detecting hybridization signals (see Chapter 1). Alternative procedures can be used which combine a longer time of digestion with smaller amounts of enzyme or vice versa. Due to the variable efficiency of the enzyme from one batch to another, it is advisable to carry out a pre-test to optimize conditions. We advise that each batch of proteinase K is tested by using a small fixed amount of enzyme (1 μg/ml digestion solution) and an increasing time of treatment at a fixed temperature of 37°C. A stock of 50–100 small aliquots is prepared from the same batch, and each aliquot is used only for a single experiment. The important parameters for hybridization are the temperature, the concentration and specific activity of probe, and hybridization time (see Chapter 1), and it is best to optimize these parameters and to use the same conditions for all probes. Although heterologous probes can be used and washed at low stringency, we find that these frequently give problems in the interpretation of faint signals, and we therefore always use homologous probes. The following procedure (*Protocol 7*) has been applied to many genes with different characteristics. However, maintaining standard conditions for both temperature and time of hybridization, the concentration and length of the probe appear to be the most important parameters. Empirically, we find that increasing the concentration of radiolabelled probe above $3-5 \times 10^4$ c.p.m. per µl of hybridization mix does not result in a proportional improvement of the signal or in much reduction in exposure time. We recommend using probe of 0.4–0.8 kb long as template and a final concentration of $3-5 \times 10^4$ c.p.m./µl of hybridization mix.

Protocol 6. Pre-treatment of sections

Reagents

- PFA, PBS, saline (see *Protocol 1*)
- Dilution buffer for proteinase K: 50 mM Tris–HCl, 5 mM EDTA pH 8—sterilize by autoclaving
- 0.1 M triethanolamine–HCl pH 8—sterilize by autoclaving
- Proteinase K: freshly dilute from a 10 mg/ml stock to a final concentration of 1 µg/ml in dilution buffer
- Acetic anhydride: this is toxic and must be dispensed in a fume-hood

Method

1. Dewax slides in xylene[a] twice for 10 min.

2. Place slides in 100% ethanol for 2–3 min to eliminate xylene.

3. Wash slides in 100% ethanol (twice), and then in decreasing concentrations of 95%, 85%, 60%, and 30% ethanol for 5 min each.

4. Transfer slides into saline solution and then PBS for 5 min each.

5. Refix slides in freshly prepared PFA for 20 min.

6. Wash twice with PBS for 5 min.

7. Transfer slides to proteinase K solution and leave at 37 °C for 30 min.

8. Wash in PBS for 5 min.

9. Repeat fixation of step 5.

10. Wash in distilled water for 30 sec and place in 0.1 M triethanolamine–HCl in a container with a rotating stir bar placed in a fume-hood.

11. Add acetic anhydride to a concentration of 0.25% (v/v) and leave for 10 min. When the acetic anhydride has fully dispersed the stirrer can be turned off.

12. Wash with PBS and saline solution for 5 min each.

Protocol 6. *Continued*

13. Dehydrate slides by passing through 30%, 60%, 85%, 95%, and 100% (twice) ethanol for 2–3 min each.

14. Air dry and use for hybridization.

[a] The volume can be adjusted according to the number of slides and to the size of the container. We try to combine experiments to obtain a final number of 25 or 50 slides. For 50 slides, 400 ml for each solution are enough. Xylene or ethanol can be used several times and stored in bottles at room temperature.

Protocol 7. Hybridization of sections

Equipment and reagents

- Sealable box for slides
- Incubator at 55 °C

- Hybridization mix: 50% deionized formamide, 0.3 M NaCl, 20 mM Tris–HCl pH 7.5, 5 mM EDTA pH 8, 10% dextran sulfate, 1 × Denhardt's solution, 0.5–1 mg/ml tRNA— a stock of 200 ml of hybridization mix should be prepared and stored at –20 °C

Method

1. Dilute the probe in the hybridization mix at a final concentration of $3–5 \times 10^4$ c.p.m./μl.

2. Heat the hybridization mix at 80 °C for 5 min and cool to room temperature.

3. Apply hybridization mix to each dried slide[a] and with a small piece of Parafilm gently distribute the mix all along the surface of the slides.

4. Place a piece of Parafilm over the slides and carefully squeeze out air bubbles.

5. Place the slide horizontally in a slide box containing tissue paper soaked in 50% formamide, 5 × SSC, then seal the humidified box.

6. Incubate at 55 °C for 16–18 h.

[a] 60–70 μl of hybridization mix is sufficient to cover the entire surface of a slide. Therefore 3–4 ml of hybridization mix is sufficient for 50 slides that are fully covered with sections.

---

6. Post-hybridization washes

Following hybridization, the unbound probe is removed by high stringency washes and RNase treatment (*Protocol 8*). When using ^{35}S-labelled probes an important precaution is to use a reducing agent such as dithiothreitol in washing solutions (not during RNase treatment).

Protocol 8. Post-hybridization washes

Equipment and reagents

- Rack for 25 slides
- Slide jar or plastic box suitable for the slide rack
- 20 × SSC: 3 M NaCl, 0.3 M sodium citrate dihydrate
- Solution 1: 5 × SSC, 15 mM DTT—pre-warm to 55°C just before use

- Solution 2: 50% formamide, 2 × SSC, 15 mM DTT—pre-warm to 65°C just before use
- NTE buffer: 0.5 mM NaCl, 10 mM Tris–HCl pH 7.5, 5 mM EDTA pH 8
- RNase solution: dilute ribonuclease A in NTE buffer to a final concentration of 20 μg/ml

Method

1. Remove the slide rack, place in 250 ml solution 1, then gently remove the Parafilm pieces covering the slides and discard them.

2. Transfer the slide rack to 250 ml solution 1 and incubate at 55°C for 1 h.

3. Transfer to solution 2 at 65°C for 1 h.

4. Wash three times for 15 min each with NTE buffer at room temperature.

5. Treat with RNase solution at 37°C for 30 min.

6. Wash with NTE buffer for 15 min at room temperature.

7. Repeat step 3.

8. Wash in 2 × SSC and then in 0.1 × SSC for 15 min each at room temperature.

9. Dehydrate slides by passing through 30%, 60%, 85%, and 95% ethanol, all including 0.3 M ammonium acetate, followed twice by 100% ethanol.

10. Air dry.

---

7. Autoradiography

Autoradiography (*Protocol 9*) is one of the most important steps, since in our experience problems with this step are the most frequent causes of failure to achieve good *in situ* hybridization results (see also Chapter 2 for other methods). Using standard conditions and precautions, a good result is guaranteed, but a small lack of attention in the autoradiography and/or developing can destroy a lot of work and time. Thus we strongly advise several important precautions:

(a) To perform all the manipulations in a double door dark-room.

(b) To always verify by dipping blank slides that the emulsion is without intrinsic background.

(c) To use appropriate safelight filters that provide sufficient yellow illumination.

Nevertheless, if after developing a couple of slides, it is found that there is a high background, there is an easy procedure that can save the experiment (see the end of *Protocol 10*).

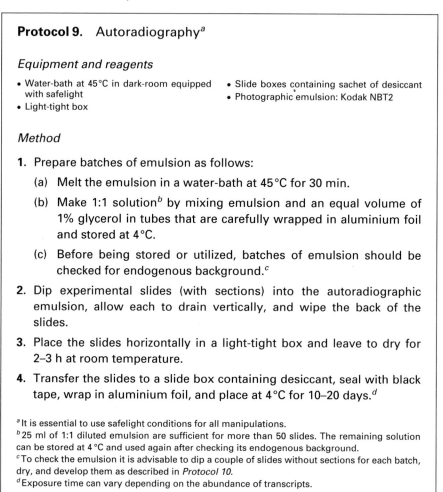

Protocol 9. Autoradiography[a]

Equipment and reagents

- Water-bath at 45°C in dark-room equipped with safelight
- Light-tight box
- Slide boxes containing sachet of desiccant
- Photographic emulsion: Kodak NBT2

Method

1. Prepare batches of emulsion as follows:
   (a) Melt the emulsion in a water-bath at 45°C for 30 min.
   (b) Make 1:1 solution[b] by mixing emulsion and an equal volume of 1% glycerol in tubes that are carefully wrapped in aluminium foil and stored at 4°C.
   (c) Before being stored or utilized, batches of emulsion should be checked for endogenous background.[c]

2. Dip experimental slides (with sections) into the autoradiographic emulsion, allow each to drain vertically, and wipe the back of the slides.

3. Place the slides horizontally in a light-tight box and leave to dry for 2–3 h at room temperature.

4. Transfer the slides to a slide box containing desiccant, seal with black tape, wrap in aluminium foil, and place at 4°C for 10–20 days.[d]

[a] It is essential to use safelight conditions for all manipulations.
[b] 25 ml of 1:1 diluted emulsion are sufficient for more than 50 slides. The remaining solution can be stored at 4°C and used again after checking its endogenous background.
[c] To check the emulsion it is advisable to dip a couple of slides without sections for each batch, dry, and develop them as described in *Protocol 10*.
[d] Exposure time can vary depending on the abundance of transcripts.

8. Developing, staining, and photography

After the appropriate period of autoradiography, the slides are developed, and the tissues stained and mounted using coverslips. Sections should be stained faintly, since it is very difficult to identify hybridization signal on darkly-stained sections. To observe the expression, sections should be first analysed at low magnification using a microscope with dark-field illumination (see also Chapter 2). Once the expression pattern is visualized, the most informative sections may then require a more detailed analysis at higher magnification. Many different types of film are suitable for photography of the images, and we find that Kodak TMax100 film gives satisfactory results for both faint and strong signals.

Protocol 10. Developing and staining

Reagents
- Developing solution: Kodak D19 developer at 160 g/litre
- Fixing solution: Kodak Unifix at 130 g/litre

A. *Developing slides*
1. Remove box with slides from 4°C and leave 1 h at room temperature.
2. In the dark-room transfer the slides to a slide rack.[a]
3. Place the slide rack in a freshly prepared developing solution for 2 min, in 1% glycerol:1% acetic acid (alternatively a wash in water is sufficient) for 1 min, then in fixing solution for 5–10 min, all at room temperature.
4. Wash slides twice in a large volume of water for 10 min each.
5. Place slides in 0.02% toluidine blue solution[b] for 5 min, then wash in water for 1–2 min.
6. Dehydrate and destain[c] slides by passing through 30%, 60%, 85%, 95% ethanol, then twice in 100% ethanol, and then air dry.
7. Mount the slides with a clean coverslip using a few drops of a mounting agent (Permount or DPX) and dry in a fume-hood.

B. *Rescuing slides from high background in the emulsion*
If a high level of background is detected in the emulsion upon checking one or two experimental slides, proceed as follows with the remaining slides.
1. Transfer the undeveloped slides to a slide rack.
2. Place the slide rack in a low concentration fixing solution (30 g/litre) for 20 min; the slides will appear completely transparent.
3. Wash four times in a large volume of $0.1 \times$ SSC for 20 min each.
4. Dehydrate slides through 30%, 60%, 85%, 95%, then twice in 100% ethanol, and then air dry.
5. Repeat autoradiography procedure (*Protocol 9*) and develop again.

[a] Before developing all the slides, check one or two of them to verify that there is no background.
[b] Alternative stains can be used such as eosin, haematoxylin/eosin, or Giemsa.
[c] To better visualize faint signals we suggest a pale blue stain.

---

9. *In situ* hybridization on cryosections

The procedure described in *Protocol 11* is particularly useful for adult tissues and very late stage embryos. The differences compared to the procedure described above are in the preparation and sectioning of the tissues and in the pre-treatment of sections before hybridization. After perfusion of the terminally anaesthetized mouse with 4% paraformaldehyde fixative, tissues or organs are dissected, cryoprotected and embedded in OCT, and then frozen in dry ice. They can then be stored at –80 °C. After sectioning, slides should be fixed and dehydrated before being hybridized or stored again at –80 °C. To increase accessibility of probe to tissues a thermal shock is carried out during the pre-treatment rather than proteinase K digestion. *Protocols 11* and *12* contain procedures for fixing, embedding, sectioning, and pre-treatment of cryostat brain sections, but they have been successfully used also for other adult tissues.

Protocol 11.   Fixation and sectioning

Reagents
- PFA: freshly prepared as described in *Protocol 1*
- Sucrose solution: 18% sucrose in PBS
- Tissue-Tek OCT medium

Method

1. Perfuse terminally anaesthetized mice with PFA by heart injection.
2. Dissect out brain and place it in PFA at 4 °C for an additional 1–2 h with gentle rotation.
3. Wash in PBS for 30 min, then place in sucrose solution, and gently rotate at 4 °C overnight.[a]
4. Transfer the brain to a mould containing Tissue-Tek OCT medium, orientate with a needle, and place in dry ice for 2 h.[b]
5. Set the cutting temperature of the cryostat to –20 °C, apply the OCT cube to the cryostat, orientate, and cut 8–10 μm sections.
6. Transfer sections one by one from the blade to the slides.
7. Dry the slides in the cryostat at –20 °C for 2 h or overnight.
8. Fix the dried slides in PFA for 3–5 min.
9. Wash briefly with PBS twice.
10. Dehydrate by passing through 30%, 60%, 85%, 95%, and 100% (twice) ethanol, and then air dry.
11. Slides can be stored at –80 °C or pre-treated for hybridization (*Protocol 12*).

[a] Steps 2 and 3 can be omitted and the brain directly transferred to the mould, but sucrose helps to prevent damage and gives better histological preservation of the tissue.
[b] At this step the mould can be stored at –80 °C.

Protocol 12. Pre-treatment of cryosections

Reagents
- PFA: freshly prepared as described in *Protocol 1*
- Solution 1: 50% formamide, 2 × SSC

Method
1. Place slides in cold acetone (4 °C) for 5 min and then air dry.
2. Fix in PFA for 15 min at 4 °C.
3. Wash slides twice in PBS for 5 min.
4. Acetylate exactly as described in *Protocol 6*, step 11.
5. Wash twice in 2 × SSC for 5 min at room temperature.
6. Place slides in pre-warmed solution 1 for 10 min at 60 °C.
7. Quickly transfer slides to pre-cooled 50% then 70% ethanol at –20 °C for 5 min.
8. Dehydrate slides in 100% ethanol twice (5 min each) at room temperature and air dry.
9. Hybridization, washing, and autoradiography are carried out exactly as described in *Protocols 7–10*.

---

10. Useful molecular markers to define gene expression patterns

The results obtained from *in situ* hybridization analyses can be of moderate or high interest depending on the ability to link experimental results to biological aspects of development and differentiation. Pioneering *in situ* expression studies described and identified crucial morphogenetic steps of development and anatomical aspects of complex tissues without the guidance of other molecular markers. Today many markers are available. Therefore, to assign the expression of a gene to a specific tissue or region it is very helpful to compare its expression pattern to that of other known genes. From such analyses potential biological and functional relationships can be proposed. To unambiguously compare different genes, it is useful to analyse a series of slides of adjacent sections. The expression pattern of many genes have been analysed by *in situ* hybridization in the mouse and other species and it is impossible to list all of them here (though see Chapter 8). However, some of them are are universally recognized as markers of important embryonic tissues at different stages of development (*Table 3*). The reader will find more details on their expression patterns in the mouse in the corresponding references.

Table 3. Markers useful to identify different tissues during embryonic development

Embryonic stage (d)	Morphogenetic events and labelled tissues	Genes and references
5.5	Pre-gastrulation	*gsc* (6); *Otx2* (7, 8)
6–7.5	Gastrulation (ectoderm, mesoderm, and endoderm)	*gsc* (6); *Otx2* (7, 8); *HNF3β* (9, 10); *Nodal* (11); *Fgf-4* (12); *T* (13); *Fgf-8* (14); *Cdx-1* (15); *Lim-1* (16, 17); *Evx-1* (18); *Kripto* (19); *Shh* (20); *Hox-B1* (21)
8–9.5	Regionalization of the rostral neural tube (future brain)	*Fgf-8* (14); *Otx2* (7, 22); *Otx1* (7, 22); *Emx2* (22, 23); *Emx1* (22, 23); *En-1* (24, 25); *En-2* (24, 25); *Pax-2* (25, 26); *Pax-5* (25, 27); *Wnt-1* (25, 28)
10 onwards	Segmental identities of the developing brain	*Fgf-8* (14); *Otx1* (7, 22); *Otx2* (7, 22); *Shh* (20); *Emx2* (22, 23); *Emx1* (22, 23); *En-1* (24, 25); *En-2* (24, 25); *Wnt-1* (25, 28); *Pax-2* (25, 32); *Pax-5* (27, 32); *Dlx-1* (29, 30); *Gbx-2* (30); *Wnt-3* (30); *Dlx-2* (30, 31); *Pax-6* (32, 33); *T-brain-1* (34); *Nkx-2.2* (35); *Otp* (36); *PLZF* (37)
8–9.5	Anteroposterior patterning of hindbrain and branchial arches	*Hox* genes (all members from paralogous groups 1–4) (2, 21, 38–42); *Krox-20* (43)
9 onwards	Anteroposterior patterning of spinal cord, somites, and vertebrae	*Hox* genes (44–46)
9 onwards	Dorsoventral patterning of the spinal cord	*Pax-2* (2, 49); *Pax-3* (47, 49); *Pax-7* (48, 49); *Pax-8* (2, 49); *Shh* (20); *Wnt-1* (25, 28); *Pax-6* (33, 49); *Otp* (36)
8 onwards	Myotome development	*Fgf-4* (12); *myf-5* (50); *myf-6* (51); *myogenin* (52); *MyoD* (52)
9 onwards	Limbs	*Fgf-4* (12); *Hox* genes (53, 54); *Fgf-8* (14, 55); *Shh* (20, 56); *msx-1* (57); *msx-2* (57); *Wnt* family (58)

Acknowledgements

I wish to thank Denis Duboule who introduced me to *in situ* hybridization. I wish also to thank Patricia Reynolds and Antonietta Secondulfo for preparing the manuscript.

References

1. Wilkinson, D. G. (ed.) (1992). *In situ hybridization: a practical approach.* IRL Press, Oxford.
2. Hogan, B., Beddington, R., Costantini, F., and Lacy, E. (ed.) (1994). *Manipulating the mouse embryo. A laboratory manual*, 2nd edn. Cold Spring Harbor Laboratory, Cold Spring Harbor, New York.

3. Theiler, K. (1989). *The house mouse: atlas of embryonic development.* Spring–Verlag, New York.
4. Rugh, R. (ed.) (1991). *The mouse: its reproduction and development.* Oxford University Press, Oxford.
5. Angerer, L. M. and Angerer, R. C. (1994). In *In situ hybridization: a practical approach* (ed. D. G. Wilkinson), pp. 15–32. IRL Press, Oxford.
6. Blum, M., Gaunt, S. J., Cho, K. W. Y., Steinbesser, H., Blumberg, B., Bittner, D., *et al.* (1992). *Cell*, **69**, 1097.
7. Simeone, A., Acampora, D., Mallamaci, A., Stornaiuolo, A., D'Apice, M. R., Nigro, V., *et al.* (1993). *EMBO J.*, **12**, 2735.
8. Ang, S.-L., Conlon, R. A., Jin, O., and Rossant, J. (1994). *Development*, **120**, 979.
9. Ang, S.-L., Wierda, A., Wong, D., Stevens, K. A., Cascio, S., Rossant, J., *et al.* (1993). *Development*, **119**, 1301.
10. Monaghan, A. P., Kaestner, K. H., Grau, E., and Schutz, G. (1993). *Development*, **119**, 567.
11. Conlon, F. L., Lyons, K. M., Takaesu, N., Barth, K. S., Kispert, A., Herrmann, B., *et al.* (1994). *Development*, **120**, 1919.
12. Niswander, L. and Martin, G. R. (1992). *Development*, **114**, 755.
13. Wilkinson, D. G., Bhatt, S., and Herrmann, B. G. (1990). *Nature*, **343**, 657.
14. Crossley, P. H. and Martin, G. R. (1995). *Development*, **121**, 439.
15. Frumkin, A., Rangini, Z., Ben-Yehuda, A., Gruenbaum, Y., and Fainsod, A. (1991). *Development*, **112**, 207.
16. Barnes, J. D., Crosby, J. L., Jones, C. M., Wright, C. V. E., and Hogan, B. L. M. (1994). *Dev. Biol.*, **161**, 168.
17. Shawlot, W. and Behringer, R. R. (1995). *Nature*, **374**, 425.
18. Dush, M. K. and Martin, G. R. (1992). *Dev. Biol.*, **151**, 273.
19. Dono, R., Scalera, L., Pacifico, F., Acampora, D., Persico, M. G., and Simeone, A. (1993). *Development*, **118**, 1157.
20. Echelard, Y., Epstein, D. J., St-Jacques, B., Shen, L., Mohler, J., McMahon, J. A., *et al.* (1993). *Cell*, **75**, 1417.
21. Frohman, M. A., Boyle, M., and Martin, G. R. (1990). *Development*, **110**, 589.
22. Simeone, A., Acampora, D., Gulisano, M., Stornaiuolo, A., and Boncinelli, E. (1992). *Nature*, **358**, 687.
23. Simeone, A., Gulisano, M., Acampora, D., Stornaiuolo, A., Rambaldi, M., and Boncinelli, E. (1992). *EMBO J.*, **11**, 2541.
24. Davis, C. A. and Joyner, A. L. (1988). *Genes Dev.*, **2**, 1736.
25. Rowitch, D. H. and McMahon, A. P. (1995). *Mech. Dev.*, **52**, 3.
26. Püschel, A. W., Westerfield, M., and Dressler, G. R. (1992). *Mech. Dev.*, **38**, 197.
27. Asano, M. and Gruss, P. (1992). *Mech. Dev.*, **39**, 29.
28. Wilkinson, D. G., Bailes, J. A., and McMahon, A. P. (1987). *Cell*, **50**, 79.
29. Price, M., Lemaistre, M., Pischetola, M., Lauro, R. D., and Duboule, D. (1991). *Nature*, **351**, 748.
30. Bulfone, A., Puelles, L., Porteus, M. H., Frohman, M. A., Martin, G. R., and Rubenstein, J. L. R. (1993). *J. Neurosci.*, **13**, 3155.
31. Porteus, M. H., Bulfone, A., Ciaranello, R. D., and Rubenstein, J. L. R. (1991). *Neuron*, **7**, 221.
32. Stoykova, A. and Gruss, P. (1994). *J. Neurosci.*, **14**, 1395.
33. Walther, C. and Gruss, P. (1991). *Development*, **113**, 1435.

34. Bulfone, A., Smiga, S. M., Shimamura, K., Peterson, A., Puelles, L., and Rubenstein, J. L. R. (1995). *Neuron*, **15**, 63.
35. Price, M., Lazzaro, D., Pohl, T., Mattei, M.-G., Rüther, U., Olivo, J.-C., *et al.* (1992). *Neuron*, **8**, 241.
36. Simeone, A., D'Apice, M. R., Nigro, V., Casanova, J., Graziani, F., Acampora, D., *et al.* (1994). *Neuron*, **13**, 83.
37. Avantaggiato, V., Pandolfi, P. P., Ruthardt, M., Hawe, N., Acampora, D., Pelicci, P. G., *et al.* (1995). *J. Neurosci.*, **15**, 4927.
38. Hunt, P., Gulisano, M., Cook, M., Sham, M.-H., Faiella, A., Wilkinson, D., *et al.* (1991). *Nature*, **353**, 861.
39. Hunt, P., Whiting, J., Nonchev, S., Sham, M., Marshall, H., Graham, A., *et al.* (1991). *Development*, **113 (Suppl. 2)**, 63.
40. Wilkinson, D. G., Bhatt, S., Cook, M., Boncinelli, E., and Krumlauf, R. (1989). *Nature*, **341**, 405.
41. Murphy, P. and Hill, R. E. (1991). *Development*, **111**, 61.
42. Duboule, D. (ed.) (1994). *Guidebook to the homeobox genes.* Oxford University Press.
43. Wilkinson, D. G., Bhatt, S., Chavier, P., Bravo, R., and Charnay, P. (1989). *Nature*, **337**, 461.
44. Graham, A., Papalopulu, N., and Krumlauf, R. (1989). *Cell*, **57**, 367.
45. Izpisúa-Belmonte, J.-C., Falkenstein, H., Dollé, P., Renucci, A., and Duboule, D. (1991). *EMBO J.*, **10**, 2279.
46. Kessel, M. and Gruss, P. (1991). *Cell*, **67**, 89.
47. Goulding, M. D., Chalepkis, G., Deutsch, U., Erselius, J. R., and Gruss, P. (1991). *EMBO J.*, **10**, 1135.
48. Jostes, B., Walther, C., and Gruss, P. (1991). *Mech. Dev.*, **33**, 27.
49. Gruss, P. and Walther, C. (1992). *Cell*, **69**, 719.
50. Ott, M.-O., Bober, E., Lyons, G., Arnold, H., and Buckingham, M. (1991). *Development*, **111**, 1097.
51. Bober, E., Lyons, G. E., Braun, T., Cossu, G., Buckingham, M. J., and Arnold, H. (1991). *J. Cell Biol.*, **113**, 1255.
52. Sassoon, D., Lyons, G., Wright, W., Lin, V., Lasar, A., Weintraub, A., *et al.* (1989). *Nature*, **341**, 303.
53. Izpisúa-Belmonte, J.-C. and Duboule, D. (1992). *Dev. Biol.*, **152**, 26.
54. Dollé, P., Izpisúa-Belmonte, J.-C., Falkelstein, H., Renucci, A., and Duboule, D. (1989). *Nature*, **342**, 767.
55. Crossley, P. H., Minowada, G., MacArthur, C. A., and Martin, G. R. (1996). *Cell*, **84**, 127.
56. Riddle, R. D., Johnson, E., Laufer, E., and Tabin, C. (1993). *Cell*, **75**, 1401.
57. Robert, B., Lyons, G., Simandl, B. K., Kuroiwa, A., and Buckingham, M. (1991). *Genes Dev.*, **5**, 2363.
58. Gavin, B. J., McMahon, J. A., and McMahon, A. P. (1990). *Genes Dev.*, **4**, 2319.

In situ hybridization of mRNA with hapten labelled probes

QILING XU and DAVID G. WILKINSON

1. Introduction

In situ hybridization with hapten labelled probes offers several advantages compared with ^{35}S-labelled radioactive probes, including a single cell resolution of the signal, the ability to directly visualize spatial patterns by hybridizing whole tissues, and the ability to visualize different transcripts simultaneously by use of multiple labelling and detection methods. Several haptenized nucleotides are available, including digoxigenin- (DIG), fluorescein-, and biotin-UTP that can be used to make labelled RNA probes. DIG and fluorescein can be detected with antibodies, and biotin with streptavidin, to which enzymes such as alkaline phosphatase (AP), β-galactosidase, or horse-radish peroxidase have been conjugated. For each of these enzymes, a variety of chromogenic substrates are available that yield different coloured products. Currently, the reagents that have found widespread favour since they give a high sensitivity and low background are DIG labelled RNA probes detected with an AP-conjugated anti-DIG antibody and the chromogenic substrate mix of 5-bromo-4-chloro-3-indolyl-phosphate (BCIP) plus 4-nitroblue tetrazolium chloride (NBT). Fluorescein labelled probes detected with an AP-conjugated anti-fluorescein antibody and BCIP/NBT give similar results. In this chapter, we describe methods for non-radioactive *in situ* hybridization of whole mount tissues and of tissue sections. For advice on controls and troubleshooting problems, see Chapter 1.

2. When to hybridize to sections or whole mounts

The major advantages of *in situ* hybridization to whole embryos compared with tissue sections are that it can easily be carried out on many samples at once and gives a direct visualization of the spatial pattern of RNA expression (see Chapter 1, *Figure 2*). As a consequence, a complete picture of the pattern of gene expression can be obtained without the need to prepare and hybridize large numbers of tissue sections. If sections are required, it is less work to section embryos after whole mount hybridization and signal detection than it is to prepare sections and then hybridize. However, whole mount hybridization does have several limitations:

(a) The limited extent of penetration of reagents into tissues restricts the size of embryos that can be used for whole mount *in situ* hybridization. Although low backgrounds can be obtained with large embryos (such as 12.5 day mouse embryos), strong signals are only obtained when the tissue is close to the surface, and thus internal sites of expression could be missed. This problem can be alleviated to some extent by longer hybridization and washing steps. Access of the reagents can be increased by bisecting embryos, or dissecting out the tissue of interest prior to hybridization.

(b) Although comparison of the expression of two genes in the same embryo can be achieved by the detection of two mRNAs in whole mount (Chapter 5), with the chromogenic substrates currently available, overlaps in expression can only be detected under optimal conditions. The hybridization of adjacent sections enables the expression patterns of many genes to be directly compared (see Chapter 3), though it is not possible to ascertain whether there is co-expression in the same cells.

The *in situ* hybridization of tissue sections is thus the method of choice for detection of gene expression in large embryos or tissues and can be the better option for the comparison of multiple genes. Typically, 6–10 μm sections are used, but a potential disadvantage is that these give weaker signals compared with whole embryos in which the signal observed is equivalent to the sum of many sections. The use of thicker sections, such as 20 μm, will give greater signal. A further alternative is to carry out *in situ* hybridization on floating sections (see also Chapter 2). This involves cutting thick wax sections (50 μm) which are then dewaxed with two 5 min washes in Histoclear, followed by two 5 min washes in methanol, and then whole mount *in situ* hybridization carried out starting at *Protocol 2*, step 5. Very strong signals are obtained by this method, but the sections can wrinkle somewhat.

3. General strategy

For those carrying out *in situ* hybridization for the first time, or with an untested probe, an important question is: what is the best strategy for getting results quickly? In our experience, the most common cause of problems is that the conditions used have given a low signal that is difficult to visualize even after major efforts to reduce background. A good starting point is therefore to use conditions that maximize the signal strength and then deal with problems with background should they occur.

(a) An important general precaution is to avoid the degradation of cellular RNAs by using ribonuclease-free solutions and containers, and wearing gloves, for all pre-hybridization steps. We find it sufficient to autoclave the PBS used for making fixative and pre-treatment solutions and to use disposable sterile plastic tubes. However, as a further precaution, ribonucleases can be inactivated by treatment of PBS with diethyl-pyrocarbonate (DEPC) (see Chapter 2, *Protocol 19*).

(b) Longer probes give stronger signals, so it is best to synthesize as long a probe as possible, which should then be size-reduced as described in *Protocol 1* if it is greater than 1 kb in length. In general, under the stringency conditions used for *in situ* hybridization, cross-hybridization to transcripts of related genes is not a problem, even when using probes corresponding to coding region sequences that are highly conserved at the amino acid level. This can be ascertained by using appropriate controls (see Chapter 1). It is important to avoid using probes that include a poly(A) tail.

(c) High stringency conditions can be used for hybridization and post-hybridization washing in order to reduce the extent of any non-specific binding of probe. However, for short or AU-rich probes, this can lead to very low signals, so we prefer to use moderate stringency conditions (such as 0.2 × SSC, 55°C) in the first instance. For most probes, low backgrounds are obtained under these conditions and it is not necessary to increase the stringency.

(d) To reduce background, non-hybridized probe is degraded in a post-hybridization RNase treatment when using radiolabelled probes, and this procedure was also used in early protocols for hapten labelled probes. However, following various improvements to these methods, it was found that this RNase treatment is not needed, and reduces the specific signal. This decrease in signal presumably is because, due to the cross-linking of tissue components and the conformation of target RNA, the probe cannot hybridize along its full-length, and thus single-stranded regions are available for attack by RNase.

(e) Non-specific binding of anti-hapten antibody can be reduced by using stronger pre-blocking conditions and/or high detergent concentrations, as used in *Protocol 4* compared with *Protocol 3*. Although the former procedure gives a lower signal, due to the clean background this is often compensated by the ability to carry out the colour development for a longer period. It may therefore be worthwhile to compare the results obtained with these two protocols.

4. Whole mount *in situ* hybridization

For whole mount *in situ* hybridization, embryos are fixed, pre-treatments carried out that increase penetration of probe, and then pre-hybridized to block non-specific binding sites. Following hybridization with hapten labelled probe, the embryos are washed to remove unbound probe, pre-blocked, and then incubated with an enzyme-conjugated antibody that binds the hapten label. After washing to remove unbound antibody, incubation with a chromogenic substrate for the enzyme generates a signal at the location of the target RNA.

For largely historical reasons, a number of distinct recipes have been developed for the *in situ* hybridization of different organisms (1–3). Although not all aspects of these protocols have been tested systematically, many of the variations appear to be unimportant. The protocols given below have been used successfully for a variety of systems, including zebrafish, *Xenopus*, chick, and mouse embryos. It is, however, worth bearing in mind that it is easier to get strong signals and low backgrounds in some systems (such as zebrafish) than others, probably due to the differences in embryo size and yolkiness of cells.

In order to obtain low backgrounds it is important that the washes are thorough, but do not damage the embryo. We use a variable speed rocking platform (Denley Reciprocal Mixer) adjusted such that the embryos are gently agitated during pre-hybridization, hybridization, and washing steps; this is easier to achieve if the tube is not completely full. For the high temperature incubations, we place microtubes in a heater block (Techne Dri-Block DB-1) turned on its side on a rocking platform. Alternatively, an incubator containing a rocking platform can be used.

4.1 Preparation of labelled RNA probe

Probe is synthesized by the *in vitro* transcription of DNA template in the presence of ribonucleotides, one of which is hapten-conjugated (e.g. with DIG or fluorescein), such that labelled RNA that is complementary to the target mRNA is produced (*Protocol 1*). The DNA template can then be removed by digesting with DNase I, but this step is optional. The RNA probe is purified away from unincorporated nucleotides, which can give high backgrounds, by ethanol precipitation in the presence of ammonium acetate or lithium chloride, or by using a gel filtration spin column.

The signal strength is related to the sequence complexity of the probe, and so, whenever possible, long probes ($\sim$ 0.5–3 kb) should be used. However, since long probes penetrate less efficiently into the tissue to obtain optimal signals the RNA probe is reduced in size by limited alkaline hydrolysis. We degrade probes of $>$ 1 kb to an average of $\sim$ 500 bp, and do not degrade probes of $<$ 1 kb, but the optimal conditions have not been systematically investigated and will depend upon the nature of the tissue and extent of cross-linking by the fixative.

The usual method to generate probe is by cloning the cDNA fragment into a vector containing T7, T3, or SP6 RNA polymerase initiation sites, such as the pGEM (Promega) or Bluescript (Stratagene) plasmid vectors. Plasmids are prepared by standard protocols, such as the alkaline lysis method. It is not necessary to use highly pure DNA, and even small scale minipreps can be used; however, if the plasmid preparation contains ribonuclease A (used to degrade bacterial RNA), this should be removed by several phenol: chloroform extractions. Probes including plasmid sequences yield high backgrounds, so to ensure that only cDNA sequences are transcribed the construct is linearized at a restriction site at the 5′ end of the cDNA. Aberrant transcripts are produced from DNA with a 3′ overhang, so this linearization should be carried out with restriction enzymes that generate either a 5′ overhang or blunt end. After checking that linearization is complete, the DNA is phenol:chloroform extracted, ethanol precipitated, then redissolved at 1 μg/μl in TE buffer.

Protocol 1. Synthesis of DIG or fluorescein labelled RNA probe

Equipment and reagents

- Water-bath, heater block, or incubator at 37°C
- Agarose gel electrophoresis apparatus
- Ultraviolet light transilluminator
- Microcentrifuge
- 1% agarose gel containing 0.5 μg/ml ethidium bromide
- 5 × transcription buffer: 200 mM Tris–HCl pH 7.9, 30 mM MgCl$_2$, 10 mM spermidine, 50 mM NaCl

- Sterile distilled water
- Nucleotide mix: 10 mM GTP, 10 mM ATP, 10 mM CTP, 6.5 mM UTP, 3.5 mM digoxigenin-UTP (or fluorescein-UTP)
- 1 μg/μl linearized plasmid in TE buffer (10 mM Tris–HCl, 0.1 mM EDTA pH 8)
- Placental ribonuclease inhibitor at 100 U/μl
- SP6, T7, or T3 RNA polymerase at 10 U/μl
- Ribonuclease-free deoxyribonuclease I (DNase I) at 1 U/μl

Method

1. Mix these reagents in the following order at room temperature:

    - sterile distilled water   9.5 μl
    - 5 × transcription buffer   4 μl
    - 0.1 M dithiothreitol   2 μl
    - nucleotide mix   2 μl
    - linearized plasmid   1 μl
    - ribonuclease inhibitor   0.5 μl
    - RNA polymerase   1 μl

2. Incubate at 37°C for 2 h.

3. Remove a 1 μl aliquot and run on an agarose gel containing 0.5 μg/ml ethidium bromide. This gel must be ribonuclease-free (we use an electrophoresis apparatus designated for this purpose, i.e. not used for plasmid minipreps that contain RNase), but does not need to be denaturing. An RNA band approx. tenfold more intense than the plasmid band should be seen on a UV transilluminator, indicating that ~ 10 μg probe has been synthesized.

Protocol 1. *Continued*

4. Add 2 μl ribonuclease-free DNase I and incubate at 37°C for 15 min. This step is optional.

5. If the transcript is 1 kb or greater in length, reduce the average size to ~ 500 bases. Add an equal volume of 80 mM $NaHCO_3$, 120 mM Na_2CO_3, mix, and heat at 60°C for a period of time (mins) = (L – 0.5)/ (0.055L), where L is the starting length (kb) of the transcript. Check the size of the product on an agarose gel, since over-degraded probes give low signals and high backgrounds.

6. (a) Either adjust volume to 50 μl, load on a microspin column (e.g. Pharmacia Biotech Microspin S-400 HR column), and follow the manufacturer's instructions. The flow-through should contain probe at ~ 0.2 μg/μl.

   (b) Or purify by ethanol precipitation. Add 130 μl dH_2O, 50 μl 10 M ammonium acetate, 400 μl ethanol, mix, and incubate at –20°C for 30 min. Do not incubate at a lower temperature as this may precipitate unincorporated nucleotides. Spin in a microcentrifuge for 10 min, wash pellet with 70% ethanol, and air dry the pellet. Redissolve in ice-cold TE at ~ 0.2 μg/μl.

7. Add an equal volume of hybridization mix (recipe in *Protocol 2*) and store at –20°C. This is stable for years. For hybridization use 1–5 μl of this stock per ml of hybridization mix.

4.2 Fixation and pre-treatment of embryos

Fixation of tissues is usually carried out using 4% paraformaldehyde. An alternative is 4% formaldehyde, and we find that this give similar results to paraformaldehyde for all of the species we have tested (*Xenopus*, chick, and mouse). A common problem with whole mount hybridizations is the trapping of reagents in enclosed cavities, such as the neural tube or heart, leading to high backgrounds. For this reason, it is important to dissect open any such cavities, for example by tearing a hole in the dorsal hindbrain. For small embryos (such as zebrafish or *Xenopus* embryos, mouse embryos up to 9 d, or chick embryos up to stage 17) a fixation time of 2 h at room temperature is adequate, but overnight fixation at 4°C gives the same results and can be more convenient.

Embryos are then dehydrated and rehydrated through a methanol series, which permeabilizes the tissues by removing lipid membranes. Further permeabilization is achieved by proteinase K treatment to partially digest cellular proteins, followed by refixation. The optimal proteinase K treatment varies according to the size of embryo, since surface tissues will be digested more rapidly than deeper tissues. *Table 1* gives a guide to the typical times used for

Table 1. Approximate times for proteinase K digestion of embryos

	Length of proteinase K digestion
Zebrafish	
Up to 20 somites	5 min
20 somites +	10 min
Xenopus	
Up to stage 14	10 min
Stage 15 +	15 min
Chick	
Stage 3–6	5 min
Stage 6–12	10 min
Stage 12–25	20 min
Mouse	
7.5 d	7 min
8.5 d	10 min
9.5 d	15 min
10.5 d	20 min
11.5 d	25 min

different systems. These times are very approximate and will depend on the batch of proteinase K; the reaction should be stopped immediately if the tissue is seen to be falling apart. If tissue disintegration occurs during subsequent steps or poor signals are obtained, the period of digestion should be optimized for the batch of proteinase K being used. The refixation step is necessary since otherwise the tissues tend to disintegrate during later steps, and a stronger fixative (0.2% glutaraldehyde, 4% paraformaldehyde) is found to be beneficial for some systems.

Protocol 2 describes the method for fixation and pre-treatment of embryos. Unless otherwise indicated, the incubations are at room temperature. The embryos are gently agitated on a rocking platform during steps 3–9 and 11. It is important to not remove all of the liquid when changing solutions, otherwise the surface tension will flatten the embryos, and they are especially delicate after the proteinase treatment. The containers used for these steps depends upon the size of the tissue and the equipment available. For small embryos, we use 2 ml microcentrifuge tubes (Eppendorf-type tubes with a snap-cap) throughout the pre-treatment, hybridization, and washing steps. For large embryos we use 7 ml flat-bottomed Bijou tubes for steps 2–10. For the pre-hybridization step, the embryos are then transferred to 2 ml microcentrifuge tubes and placed in a heater block turned on its side on a rocking platform. Several different recipes for the hybridization system have been used for different systems. Below we give two alternative recipes: one that works nicely for mouse, chick, and *Xenopus* embryos, and a simpler recipe that gives excellent results for zebrafish embryos.

Protocol 2. Fixation and pre-treatment of embryos

Equipment and reagents

- 7 ml Bijou tubes and/or 2 ml microcentrifuge tubes (with snap-cap)
- Rocking platform
- Heater block with holder for 2 ml microcentrifuge tubes
- Phosphate-buffered saline (PBS): prepare using Dulbecco 'A' tablets (Oxoid)
- Paraformaldehyde fixative (4% paraformaldehyde in PBS) is prepared by dissolving at 65°C and then cooling on ice. Caution: paraformaldehyde fumes are toxic. Can be stored for months in aliquots at –20°C.
- PBT: PBS, 0.1% Triton X-100
- Methanol
- Glutaraldehyde: 25% stock solution (Sigma)

- Proteinase K: 10 mg/ml stock solution in dH_2O—stored in aliquots at –20°C, and diluted just before use
- Hybridization solution for mouse, chick, or *Xenopus* embryos: 50% formamide, 5 × SSC, 2% blocking powder (Boehringer, 1096176; dissolve directly in this mix), 0.1% Triton X-100, 0.1% CHAPS (Sigma), 1 mg/ml tRNA, 5 mM EDTA, 50 μg/ml heparin
- Hybridization solution for zebrafish embryos: 50% formamide, 5 × SSC pH 6 (pH of stock is adjusted with 1 M citric acid), 0.1% Triton X-100, 50 μg/ml yeast RNA, 50 μg/ml heparin
- 20 × SSC stock solution: 3 M NaCl, 0.3 M sodium citrate pH 7

Method

1. Dissect embryos free of any extra-embryonic membranes.

2. Incubate overnight in paraformaldehyde fixative at 4°C.

3. Rinse the embryos with ice-cold PBT, twice for 5 min.

4. Dehydrate the embryos by washing for 10 min at each step in a graded methanol series diluted in PBT (25% methanol, 50% methanol, 75% methanol) and then twice with 100% methanol. The embryos can now be stored at –20°C for at least several months, and are stable for at least several days at room temperature.

5. Rehydrate the embryos by washing for 10 min at each step in the graded methanol series in PBT (75% methanol, 50% methanol, 25% methanol) and then twice in PBT.

6. Treat the embryos with 10 μg/ml proteinase K in PBT for 5–25 min at room temperature. The length of treatment depends upon the type of embryo and the size of the embryos (*Table 1*), and should be optimized for each batch of proteinase.

7. Rinse the embryos for 5 min with PBT.

8. Refix the embryos with fresh 0.2% glutaraldehyde, 4% paraformaldehyde in PBT for 20 min. For zebrafish embryos, 4% paraformaldehyde in PBT is used.

9. Rinse the embryos three times for 5 min with PBT.

10. Remove most of the liquid, add 0.5–1 ml hybridization solution, and allow the embryos to sink.

11. Replace with fresh hybridization solution and incubate at 55–65°C for 2 h to overnight. The volume used will depend upon the size and

number of embryos; typically 0.5–1 ml. This can be achieved by incubating in 2 ml microtubes in a heater block. The embryos can then be stored at –20°C for at least several weeks, but lower signals are obtained after more prolonged storage.

4.3 Hybridization, washing, and detection of probe

Embryos are incubated with DIG labelled probe, the unhybridized probe removed by washing at moderate stringency, and then non-specific binding sites blocked with sheep serum. The bound probe is detected with an AP-conjugated anti-DIG antibody, washed, and incubated with a chromogenic substrate for alkaline phosphatase. This produces a coloured precipitate at the sites of the target RNA.

For hybridization, probe is used at 0.1–0.5 µg (equivalent to 1–5 µl of purified probe) per ml of hybridization mix, but if high backgrounds are obtained it can be beneficial to decrease the probe concentration. Incubations of up to three days can be used for the hybridization without detriment, and may enable penetration of probe into large embryos. Hybridization and washing can be carried out between 55–65°C, and the washing solutions should be pre-warmed. For many probes, 55°C is best since stronger signals are obtained, and this can be especially important for probes that are short ($\sim$ 200 bp) or AU-rich, but if the probe gives a background problem, a higher temperature should be tested. In addition, in cases where the probe might hybridize to transcripts of related genes, this can be avoided by higher stringency hybridization or washing. It is especially important that the embryos are agitated sufficiently to ensure thorough washing after hybridization and incubation with antibody, and it may be beneficial to increase the number and/or length of washes, especially for large embryos.

No post-hybridization ribonuclease treatment to degrade unhybridized probe is carried out, as this step decreases the specific signal, and omission gives a major increase in sensitivity. However, if there are problems with high backgrounds then it is worth testing whether ribonuclease treatment has a beneficial effect. This is carried out after step 4, by washing embryos in 100 µg/ml RNase A in KTBT (*Protocol 3*) or MABT (*Protocol 4*) at 37°C for 30 min, then washing in buffer lacking RNase three times for 5 min. The procedure is then continued from step 5. Pre-absorption of the antibody with embryo powder, once used in many protocols, in general appears to be unnecessary. However, if you wish to check whether pre-absorption is beneficial, this is carried out as described in *Protocol 5*.

Two alternative methods for the post-hybridization steps are described below. The method in *Protocol 3* gives an excellent sensitivity, but sometimes a non-specific background can limit the detection of weak signals. CHAPS is a detergent that structurally resembles DIG, and might therefore block non-specific binding of probe. It is used in a number of protocols (derived from ref. 1), but is not required for zebrafish embryos. PBT or KTBT can be used for the binding and washing of the antibody. We routinely use the former for zebrafish embryos and the latter for other systems, but these are likely to give similar results. In some protocols, a higher concentration of detergent is used (1% Triton X-100); this gives cleaner backgrounds, but can substantially reduce the signal, so should only be tested if there are problems with a non-specific background.

Protocol 4 (developed by Domingos Henrique and David Ish-Horowicz) uses a stronger pre-blocking procedure prior to the detection of probe that gives a very low background. The rate of signal development is slower with *Protocol 4* compared with *Protocol 3*, but this can be compensated for by the ability to carry out the colour reaction for a longer period. In our hands, there is no advantage to using *Protocol 4* for the zebrafish embryo, but we often use it to advantage for mouse, chick, and *Xenopus* embryos.

A variety of chromogenic substrates are available for alkaline phosphatase (enabling the double detection of mRNAs—see Chapter 5), but when only one colour reaction is required NBT/BCIP is generally used since it is the most sensitive and gives a low background. An alternative is BM Purple (Boehringer, 1442074), a ready-made substrate solution that gives a very intense colour and low background. We find that this substrate is generally not suitable for whole mounts since only the surface layers of the tissue are stained, but it may be useful for chromogenic reactions on sections.

Protocol 3. Hybridization, washing, and detection of probe (method 1)

Equipment and reagents

- Shaking platform and heater block, or shaking platform in 55–65°C incubator
- Glass embryo dishes (BDH)
- Sheep serum. It may be beneficial to heat inactivate the serum at 56°C for 30 min. To avoid high backgrounds, it is important that it is free of endogenous phosphatases. Once a good batch has been identified, store in aliquots at –20°C.
- 20 × SSC stock solution
- 10% CHAPS stock solution (store at –20°C)
- Washing solution 1 (for zebrafish embryos): 2 × SSC, 0.01% Triton X-100
- Washing solution 1 (for other embryos): 2 × SSC, 0.1% CHAPS
- Washing solution 1 containing 25% formamide

- Washing solution 2 (for zebrafish embryos): 0.2 × SSC, 0.01% Triton X-100
- Washing solution 2 (for other embryos): 0.2 × SSC, 0.1% CHAPS
- PBT (see *Protocol 2*)
- KTBT: 50 mM Tris–HCl pH 7.5, 150 mM NaCl, 10 mM KCl, 0.1% Triton X-100
- NTMT: 100 mM Tris–HCl pH 9.5, 50 mM MgCl$_2$, 100 mM NaCl, 0.1% Triton X-100
- AP-conjugated anti-DIG or anti-fluorescein antibody (Boehringer)
- NBT stock solution: 75 mg/ml NBT (Boehringer) in 70% dimethylformamide
- Paraformaldehyde fixative (see *Protocol 2*)
- BCIP stock solution: 50 mg/ml BCIP (Boehringer) in dimethylformamide

Method

1. Incubate embryos with hybridization mix containing probe overnight at 55–65°C.

2. Wash in washing solution 1 containing 25% formamide, three times for 10 min, at 55–65 °C.

3. Wash in washing solution 1 for 20 min at 55–65 °C.

4. Wash with washing solution 2, three times for 20 min, at 55–65 °C.

5. Rinse with KTBT twice for 10 min each at room temperature. Alternatively use PBT in place of KTBT for this and all subsequent steps.

6. Pre-block the embryos with 10% sheep serum in KTBT for 1–3 h at room temperature.

7. Incubate with 1/2000 dilution of anti-DIG or anti-fluorescein antibody (which can be pre-absorbed with embryo powder as described in *Protocol 5*) in 10% sheep serum in KTBT, and rock overnight at 4 °C. The antibody solution can be recovered and used up to three times in total; for storage the solution must contain 0.01% sodium azide.

8. Wash the embryos with KTBT five or more times for 1 h each at room temperature, then overnight at 4 °C. The overnight wash is optional, but gives lower backgrounds (especially for larger embryos), and it is more convenient to monitor a slow colour reaction if it is started on the following morning.

9. Wash twice in NTMT for 15 min and transfer to a glass embryo dish for easier observation; crystals can form if plastic Petri dishes are used. The NTMT can include 1 mM levamisol, which inhibits many endogenous alkaline phosphatases, but for many systems this is not needed.

10. Incubate in the dark with NTMT containing 4.5 μl/ml NBT stock solution, 3.5 μl/ml BCIP stock solution. Periodically monitor the reaction and when a strong signal is produced and/or any background is observed, stop by washing several times with PBT. Fix in 4% paraformaldehyde in PBS for 1–2 h at room temperature, and store in PBS at 4 °C.

Protocol 4. Hybridization, washing, and detection of probe (method 2)

Reagents

- Washing solution: 50% formamide, 1 × SSC, 0.1% Tween 20
- Sheep serum (see *Protocol 3*)
- MABT: 100 mM maleic acid pH 7.5, 150 mM NaCl, 0.1% Tween 20 (pH is adjusted using NaOH)
- NTMT: 100 mM Tris–HCl pH 9.5, 50 mM MgCl$_2$, 100 mM NaCl, 0.1% Triton X-100
- Paraformaldehyde fixative (see *Protocol 2*)

- NBT stock solution: 75 mg/ml NBT (Boehringer) in 70% dimethylformamide
- BCIP stock solution: 50 mg/ml BCIP (Boehringer) in dimethylformamide
- Blocking powder (Boehringer, 1096176): make 10% stock by dissolving in MABT at 65 °C
- AP-conjugated anti-DIG or anti-fluorescein antibody (Boehringer)

Protocol 4. *Continued*

Method

1. Incubate embryos with hybridization mix containing probe overnight at 55–65°C.

2. Wash with washing solution, twice for 30 min, at 55–65°C.

3. Wash with 1:1 washing solution:MABT for 10 min at 55–65°C.

4. Wash with MABT three times for 5 min, then twice for 15 min, at room temperature.

5. Pre-block the embryos with 2% blocking powder, 20% sheep serum in MABT, 2–4 h at room temperature.

6. Incubate with 1/2000 dilution of anti-DIG or anti-fluorescein antibody (which can be pre-absorbed with embryo powder as described in *Protocol 5*) in 10% sheep serum in MABT, and rock overnight at 4°C. The antibody solution can be recovered and used up to three times in total.

7. Wash the embryos for 1 h with MABT at room temperature, five or more times, then overnight at 4°C. The overnight wash is optional, but gives lower backgrounds, and it is more convenient to monitor a slow colour reaction if it is started on the following morning.

8. Wash twice in NTMT for 15 min and transfer to a glass embryo dish for easier observation; crystals can form if plastic Petri dishes are used. The NTMT can contain 1 mM levamisol, which inhibits many endogenous alkaline phosphatases, but for many systems this is not needed.

9. Incubate in the dark with NTMT containing 4.5 µl/ml NBT stock solution, 3.5 µl/ml BCIP stock solution. Periodically monitor the reaction and when a strong signal is produced and/or any background is observed, stop by washing several times with PBT. Fix in 4% paraformaldehyde in PBS for 1–2 h at room temperature, and store in PBS at 4°C.

If the embryos are overstained, it is possible to partially destain them in PBS containing 1% Triton X-100, or in methanol, but care has to be taken that these do not remove too much signal. For reasons that are unclear, sometimes the colour reaction yields a brown product, and washing in PBT or briefly in methanol will convert this to a blue colour.

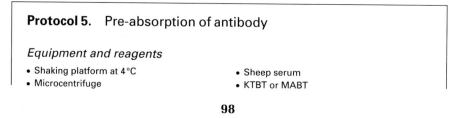

Protocol 5. Pre-absorption of antibody

Equipment and reagents
- Shaking platform at 4°C
- Microcentrifuge
- Sheep serum
- KTBT or MABT

- Embryo powder: should if possible be prepared from the same species as you are carrying out *in situ* hybridization on, but does not need to be from the same developmental stage (more material can be prepared from later stage embryos)

- AP-conjugated anti-DIG or anti-fluorescein antibody

A. *Preparation of embryo powder*

1. Homogenize embryos in a minimum volume of PBS.

2. Add 4 vol. of ice-cold acetone, mix, and incubate on ice for 30 min.

3. Centrifuge at 10 000 *g* for 10 min and remove supernatant.

4. Wash pellet with ice-cold acetone and centrifuge again.

5. Spread the pellet out and grind it into a fine powder on a sheet of filter paper.

6. Air dry and store in an airtight tube at 4 °C. Is stable for years.

B. *Pre-absorption of antibody*

1. For 2 ml of final solution, place ~ 3 mg embryo powder in a microtube and add 0.5 ml 10% sheep serum in KTBT (if using *Protocol 3*) or MABT (if using *Protocol 4*) and 1 μl antibody.

2. Shake tube (lying on its side) gently at 4 °C for at least 2 h.

3. Spin for 1 min in a microcentrifuge, remove the supernatant, and dilute it to 2 ml in buffer. The pre-absorbed antibody is stable at 4 °C.

5. *In situ* hybridization to tissue sections or cultured cells

The major steps for *in situ* hybridization of sections or cells on slides are identical (probe preparation—*Protocol 1*) or similar (pre-treatments, hybridization, washing, and detection) to the procedure for whole embryos. Since the penetration of reagents into sections is more efficient than for whole tissues, it may be possible to further reduce the length of the binding and washing steps.

5.1 General precautions

Precautions must be taken to avoid ribonucleases degrading cellular RNA prior to hybridization. PBS used for all steps prior to hybridization must be autoclaved, and all containers and slide holders must be clean and ribonuclease-free. Gloves must be worn for handling slides, and do not use any container that has been used for the ribonuclease treatment of tissues.

5.2 Preparation of tissue sections or cultured cells on slides

Tissue sections are prepared either by embedding fixed embryos in paraffin wax and cutting sections on a microtome (*Protocol 6*A), or by cutting sections of frozen tissue using a cryostat (*Protocol 6*B). A number of alternative solvents can be used for embedding tissue in wax. Xylene penetrates the tissue efficiently, but care has to be taken not to incubate the tissues for too long as this will make them brittle. Longer incubations are required for toluene, but this solvent has less of a hardening effect. These solvents must be used in a fume-hood. We routinely use Histoclear which has the advantage of being non-toxic. Commonly, 6–10 μm sections are cut, but to maximize signal it may be advantageous to cut thicker sections of up to 20 μm. The sections are dried onto slides that have been subbed (*Protocol 6*C) to promote sticking of the tissue, and can be stored prior to pre-treatment and hybridization.

For cultured cells, seed them onto slides that have been coated with poly-lysine or laminin (see also Chapter 2). Once the cells have reached the appropriate density, dehydrate them by passing quickly through 25%, 50%, 75%, and 100% ethanol, then air dry, and store in an airtight container at –70 °C.

Protocol 6.   Preparation of sections

Equipment and reagents

- For wax sectioning: microtome, graded methanol series in PBS, Histoclear (National Diagnostics), paraffin wax (melting temperature 56–58 °C, e.g. Fibrowax, BDH 36142), incubator or wax embedding apparatus at 60 °C, bath for floating sections or slide warmer at 50 °C
- For cryostat sectioning: cryostat, OCT medium (BDH), 0.5% low melting temperature agarose, 5% sucrose in PBS, 30% sucrose in PBS

- Microscope slides (preferably with a frosted end so can be labelled with pencil)
- Glass vials
- Glass embryo dishes
- 4% paraformaldehyde in PBS (see *Protocol 2*)
- TESPA (3-aminopropyltriethoxysilane, Sigma)
- Acetone

A. *Embedding in paraffin wax and sectioning*

1. Dissect the tissue or embryos and fix them in 4% paraformaldehyde in PBS, overnight at 4 °C.

2. Wash with PBS, twice for 10 min.

3. Dehydrate by taking through methanol series in PBS (25% methanol, 50% methanol, then 75% methanol), then twice in 100% methanol, for 10 min each. Large embryos or tissues should be washed for longer to ensure complete dehydration. It is convenient to carry these and the subsequent steps out in sterile glass vials.

4. Equilibrate with Histoclear, three times for 20 min, then with molten paraffin wax at 60 °C, three times for 20 min, occasionally agitating the vial.

5. Transfer the embryos to glass embryo dishes (pre-heated to 60°C), ori-entate them with a warmed hypodermic needle under a dissection microscope, and allow the wax to set. Paraffin wax blocks can be stored indefinitely at 4°C until required for use.

6. Trim the block, mount on the microtome (the way this is done will depend on the apparatus), and cut 6–10 μm sections as ribbons.

7. Float a suitable length (that will fit onto a slide) of the ribbon on a bath of distilled water at 50°C until the creases disappear, and collect on TESPA subbed slides (*Protocol 6C*). If a slide-bath is not available, can lay slides on slide warmer at 50°C, pipette on a pool of distilled water, and then float the ribbon of sections on this until the creases have disappeared.

8. Drain the excess liquid, then dry the sections onto the slides at 37°C overnight. They can be stored desiccated at 4°C.

B. Cryostat sectioning

1. (a) For small tissues: fix in 4% paraformaldehyde in PBS overnight at 4°C. To facilitate orientation of very small tissues, can then embed in 0.5% low melting temperature agarose, 5% sucrose in PBS. Equilibrate in 30% sucrose in PBS overnight at 4°C.

   (b) For large tissues: fix, then equilibrate with 30% sucrose (as above), or freeze directly in liquid nitrogen.

2. Use OCT medium to glue the tissue or agarose block onto the cryostat holder (pre-chilled to –20 to –30°C).

3. Cut sections onto TESPA subbed slides (*Protocol 6C*).

4. Air dry, and store in airtight container at –70°C.

C. Preparation of subbed slides

1. Dip the slides in 2% TESPA in acetone for 10 sec.

2. Rinse twice with acetone, and then with distilled water.

3. Dry at 37°C.

5.3 Pre-treatments, hybridization, washing, and detection of probe

Prior to hybridization, the sections are dewaxed, permeabilized by proteinase K treatment, then refixed, and dehydrated (*Protocol 7*). The probe is spread over the dry sections under a coverslip. The same general considerations apply as for whole mount hybridizations (Sections 4.2 and 4.3), with the additional factor that over-digestion with proteinase can lead to the sections falling off the slides. It is important to ascertain the optimal period of digestion with proteinase K that gives maximal signal. Unlike for whole mount hybridization, a pre-hybridization blocking step is not carried out. Instead, the hybridization mix is placed directly on the dried section and overlaid with a coverslip (some workers use Parafilm instead—see Chapter 3). The hybridization, washing, and detection of probe are carried out as described in *Protocol 8*.

The slides are placed in holders suitable for 250 ml slide dishes and 200–250 ml of solution used for *Protocol 7*, steps 1 and 3–6, and for *Protocol 8*, steps 3–7 and 11. If only a few slides are being processed, Coplin jars can be used instead. *Protocol 8*, steps 8, 10, 12, and 13 are carried out by overlaying the sections with the reagent and incubating in a humid chamber.

Protocol 7. Pre-treatment of sections or cells on slides

Equipment and reagents
- Slide racks and appropriate containers (e.g. 250 ml plastic boxes or glass dishes)
- 4% paraformaldehyde in PBS (see *Protocol 2*)
- PBS
- 10 μg/ml proteinase K: freshly diluted in PBS from a 10 mg/ml stock in distilled water

Method

1. (a) For dehydrated cryostat sections or cells on slides: warm up to room temperature. Fix in 4% paraformaldehyde in PBS for 20 min. Wash three times for 5 min each in PBS.

   (b) For wax sections: dewax the slides in Histoclear, twice for 10 min. Wash in 100% methanol for 2 min to remove most of the Histoclear. Transfer the slides through 100% methanol (twice), 75%, 50%, and 25% methanol in PBS, for 1–2 min in each solution. Wash twice in PBS for 5 min each. Optional: immerse the slides in fresh 4% paraformaldehyde in PBS for 20 min, then three times for 5 min in PBS.

2. Drain the slides and place horizontally on a clean sheet of absorbent paper on the bench. Overlay the sections with 10 μg/ml proteinase K and leave for 5–10 min.

3. Shake off excess liquid and wash the slides with PBS for 5 min.

4. Fix in fresh 4% paraformaldehyde in PBS for 20 min.

5. Wash the slides three times with PBS for 5 min.

6. Dehydrate by passing through 25%, 50%, 75% methanol in PBS, then twice in 100% methanol, for 1–2 min in each solution.

7. Air dry. Should be used on the same day for hybridization (though if required, can be stored in airtight container at –20°C for at least several days).

Protocol 8. Hybridization, washing, and detection of probe

Equipment and reagents

- Slide racks and containers
- Incubator at 55–65 °C
- 50% formamide, 2 × SSC
- 25% formamide, 2 × SSC
- 2 × SSC
- 0.2 × SSC

- PBT
- AP-conjugated anti-DIG antibody (can be pre-absorbed)
- NTMT, BCIP stock solution, NBT stock solution (see *Protocol 3*)

Method

1. Apply the hybridization mix containing probe (*Protocol 1*) to the slide adjacent to the sections ($\sim$ 5 μl/cm^2 of coverslip is sufficient) and gently lower a clean coverslip so that the mix is spread over the sections without trapping air bubbles.

2. Place the slides horizontally in a box containing tissue paper soaked in 50% formamide, 5 × SSC, seal the box, and incubate overnight at 55–65 °C.

3. Place the slides in a slide rack and immerse in pre-warmed 25% formamide, 2 × SSC at 55–65 °C until the coverslips fall off.

4. Wash with 25% formamide, 2 × SSC for 30 min at 55–65 °C.

5. Optional: rinse twice with 2 × SSC and treat with 20 μg/ml RNase in PBS at 37 °C for 30 min. This step will decrease the signal so should only be carried out if required to reduce background or to prove specificity.

6. Wash with 2 × SSC, then twice in 0.2 × SSC (30 min each step) at 55–65 °C.

7. Wash with PBT, three times for 10 min, at room temperature.

8. Quickly drain each slide and place horizontally in a sandwich box containing moist tissue paper. Take care that the sections do not become dry, and quickly overlay them with 5% sheep serum in PBT. Seal the box, and incubate for 30 min.

9. If desired, the antibody can be pre-absorbed as described in *Protocol 5*.

10. Remove the 5% serum from the sections, replace with 1/2000 diluted AP-conjugated anti-DIG antibody in 5% sheep serum, PBT and incubate in a moist box at room temperature for 1–3 h, or at 4 °C overnight.

11. Wash with PBT twice for 5 min and then three times for 15 min.

12. Wash with NTMT, three times for 5 min.

Protocol 8. *Continued*

13. Incubate in the dark with NTMT containing 4.5 µl NBT, 3.5 µl BCIP per ml. Occasionally monitor under dissecting microscope, and when sufficient signal has developed, stop the colour reaction by washing with PBT.

14. Fix the signal by immersing the slides in 4% paraformaldehyde in PBS for 30 min, then mount under a coverslip using 70% glycerol. Alternatively, for permanent mounting, fix for several hours or overnight, dehydrate quickly through a graded methanol series followed by Histoclear, then mount under a coverslip using Permount (Sigma) or DPX (BDH) mounting agent.

6. Photography

Photography of whole embryos can be carried out using a dissection microscope fitted with a camera, illuminating the specimen from above and/or with dark-field from below. The embryos can be partially cleared by equilibrating in 70% glycerol in PBT, and photographed in a Petri dish. Sometimes, a plane of view is required that cannot be obtained with the embryos resting on a flat surface, and this can be achieved as follows:

(a) Pour a ~ 2–3 mm layer of 1% agarose into a Petri dish and allow it to set.

(b) Troughs that can accommodate the embryos are then made in the agarose, either by cutting with a scalpel blade or by melting with a hot glass capillary pipette.

(c) Overlay the agarose with PBT, then orientate the embryo in a trough as required.

Higher power photographs can be shot by mounting the embryos under a coverslip. An easy way to achieve this is to place two blobs of petroleum jelly about 1 cm apart on a microscope slide, pipette the embryo (equilibrated with 70% glycerol in PBT) between these and orientate as desired, then lower a coverslip on top, gently pushing it down until the tissue is secured in place. In some cases, it can be very useful to partially dissect the embryo in order that the tissue can be flattened during mounting such that expressing cells are in the same plane of focus (see Chapter 1, *Figure 2*). To visualize cells, a light counterstaining with eosin can be carried out, but since this can obscure the signal it is preferable to use differential interference contrast (DIC, Nomarski) optics.

It can be very informative to observe gene expression in sectioned tissue, either after hybridization of sections or the sectioning of whole mount hybridized embryos. For the latter, signals can be detected in sectioned material if they are strong in the whole mount; if the signal is weak it is advisable to cut thick sections. Sections can be prepared on a cryostat or after embedding in wax, and then mounted under a coverslip (*Protocol 9*). Embedding in wax can only be carried out if the signal is intense, as some of the signal is dissolved during the incubations in organic solvents.

Protocol 9. Sectioning of whole mount hybridized embryos

Reagents
- See *Protocol 6*

A. *Cryostat sectioning*

1. Fix the stained embryos overnight in 4% paraformaldehyde in PBS.

2. Wash several times in PBS, then equilibrate overnight with 30% sucrose in PBS.

3. Mount on a cryostat chuck with OCT compound and freeze on dry ice.

4. Cut 10–25 μm sections on a cryostat.

5. Mount under a coverslip with 70% glycerol in PBS.

B. *Wax sections*

1. Fix the stained embryos overnight in 4% paraformaldehyde in PBS. This step is especially important for wax embedding because otherwise the NBT/BCIP reaction product will dissolve in the solvents used.

2. Wash the embryos twice for 10 min with PBS, then replace solution with 75% then 100% methanol (twice) for 10 min each, and with Histoclear three times for 10 min each. Shake gently to ensure efficient equilibration. These washes will need to be longer for large tissues (e.g. > 9.5 d mouse embryos), but should not be excessively prolonged.

3. Replace the solvent with three changes of paraffin wax at 60°C for 15 min each, with occasional shaking to mix. The length of these incubations may need to be adjusted according to the size of the tissue.

4. Transfer to an embryo dish at 60°C, place at room temperature, orientate using a warmed needle (if necessary, observing with a dissection microscope), and let the wax set.

5. Cut sections, mount on subbed slides (*Protocol 6C*), and dry at 37°C overnight.

6. Dewax for 2–5 min with Histoclear.

7. While the slide is still wet, mount the sections under a coverslip using Permount (Sigma) mounting agent.

We photograph with Kodak EPY64T colour reversal (slide) film, which can be used directly for slides, for preparing prints, or scanned to make a digital image. Typically, an exposure equivalent to ASA 32 (e.g. ASA64 + 1 stop) gives the best results when using DIC optics with this film, but it may be necessary to try several different exposures. Colour negative film is cheaper to print from, but it can be difficult to obtain a consistent colour balance if using commercial printers. For black and white photography, a film such as Kodak TMax 100 is suitable. Excellent advice on how to take good photographs is given in ref. 4.

Acknowledgements

We are grateful to Wendy Hatton for advice on sectioning, to Domingos Henrique, Trevor Jowett, Jeff Christiansen, and Vicky Robinson for discussions and protocols, and to the participants of EMBO *in situ* hybridization courses for testing many parameters.

References

1. Harland, R. M. (1991). In *Methods in cell biology* (ed. B. K. Kay and H. B. Peng), Vol. 36, p. 685. Academic Press, San Diego.
2. Conlon, R. A. and Herrmann, B. G. (1993). In *Methods in enzymology* (ed. P. M. Wassarman and M. L. DePamphilis), Vol. 225, p. 373. Academic Press, San Diego.
3. Wilkinson, D. G. (1992). In *In situ hybridization: a practical approach* (ed. D. G. Wilkinson), 1st edn, p. 75. IRL Press, Oxford.
4. Stern, C. D. (1993). In *Essential developmental biology: a practical approach* (ed. C. D. Stern and P. W. H. Holland), p. 67. IRL Press, Oxford.

5

Two colour *in situ* hybridization

T. JOWETT

1. Methods of localization of multiple transcripts

Once the expression pattern of a gene has been established it is often necessary to relate it to that of other genes expressed at the same period of development. Transcripts from different genes may be expressed in complementary or overlapping domains. The temporal and spatial pattern of expression of different genes can be related by performing *in situ* hybridization on consecutive sections of tissue but it is also possible to perform hybridizations with two probes and to visualize the signals separately in the same tissue.

To perform double label experiments, the antisense probes must be differentially labelled to allow different methods of detection. Antisense probes are synthesized with either digoxigenin-11-UTP or fluorescein-12-UTP. Biotinylated probes may be used in hybridizations to *Drosophila* embryos but they give high backgrounds with vertebrate embryos and tissues. Fluorescein and digoxigenin labelled probes are detected with antibodies which are conjugated with either alkaline phosphatase or horse-radish peroxidase. It is most convenient to mix the antisense probes and perform the hybridizations together. The unhybridized probes are washed off and the tissues incubated in blocking solution to prevent non-specific binding of the antibodies. The signals are then visualized by staining with appropriate enzyme substrates which produce coloured, insoluble precipitates.

Before attempting double *in situ* hybridizations it is advisable to perform separate *in situ* hybridizations with single probes labelled with digoxigenin and fluorescein for each antisense RNA. This will indicate which is the best strategy to use for each particular transcript. Some target transcripts may be significantly less abundant and will require the most sensitive method. If the results indicate that the two signals may overlap then this should be taken into consideration when deciding which method of two colour *in situ* hybridization to use.

The procedures for probe synthesis, tissue fixation, and pre-treatments before hybridization are the same as when using a single probe. Appropriate protocols are described in Chapter 4. The hybridization buffer should have a pH of 6.5, as fluorescein labelled probes appear to be less stable at acidic pH. Three suitable hybridization solutions are described in *Table 1*: dHybe

Table 1. General reagents and solutions

100 × **Denhardt's solution**: 2% (w/v) Ficoll, 2% polyvinylpyrrolidone (PVP), 2% bovine serum albumin (BSA) in water. Store in aliquots at –20 °C.

20 × **SSC** (for hybridization and washing): 3 M NaCl, 300 mM trisodium citrate. Dissolve 175.3 g NaCl and 88.2 g sodium citrate in 800 ml water. Adjust pH with 1 M citric acid to 4.5 or 6 depending on the hybridization solution used. Adjust volume to 1 litre and sterilize by autoclaving.

AP inactivation buffer: 0.1 M glycine–HCl pH 2.2, 0.1% Tween 20.

BCIP (5-bromo-4-chloro-3-indolyl-phosphate also known as X-phosphate 4-toluidine salt; Boehringer, 760 994): dissolve at 50 mg/ml in dimethylformamide. Store in aliquots at –20 °C.

Blocking solution for antibodies: 1 × PBS, 0.1% Tween 20, 2 mg/ml BSA (BDH, 44155), 5% sheep serum (Gibco BRL, 035–6070H), 1% dimethyl sulfoxide DMSO (Merck-BDH, 28216).

CHAPS (Sigma, C 3023): 10% (w/v) stock solution in deionized water. Store in aliquots at –20 °C.

DAB staining solution: 0.5 mg/ml diaminobenzidine in PBT. DAB is a potent carcinogen. All contaminated disposable materials should be incinerated and glassware soaked in 6% sodium hypochlorite solution.

dHybe: 50% formamide, 5 × SSC, 100 µg/ml yeast RNA, 0.1% Tween 20, 50 µg/ml heparin, adjust to pH 6.5 with 1 M citric acid.

ELF™ stop reaction buffer: 25 mM EDTA, 0.05% Triton X-100 in PBS pH 7.2. Dissolve EDTA in PBS and check pH.

ELF™–AP substrate kit (Molecular Probes Inc., E-6601): dilute the substrate 1:20 in ELF™ Reaction Buffer supplied with the kit. Use within 30 min. Tissues should be equilibrated with pre-reaction wash buffer (30 mM Tris, 150 mM NaCl pH 7.5) prior to adding the diluted substrate solution. The reaction is stopped with 25 mM EDTA, 0.05% Triton X-100 in PBS; the final pH should be 7.2 (addition of 1 mM levamisole is optional).

Fast Red tablets (alkaline phosphatase substrate; Boehringer, 1 496 549): each tablet contains 0.5 mg naphthol substrate, 2 mg Fast Red chromogen, and 0.4 mg levamisole. Store tablets at –20 °C. Wear gloves and use plastic forceps to handle the tablets. Dissolve one tablet in 2 ml of 100 mM Tris–HCl pH 8.2. Use the solution within 30 min.

Goat anti-mouse (IgM)–horse-radish peroxidase (Sigma, A 8786).

Goat anti-rabbit IgG (H + L) (Jackson ImmunoResearch Laboratories, 111–001–003).

Goat anti-rabbit IgG (H + L) (Sigma, R 2204).

Heparin (Sigma, H 9399): make a stock solution of 100 mg/ml in deionized water. Store in aliquots at –20 °C.

MABT: 100 mM maleic acid (Sigma, M 0375), 150 mM NaCl, 0.1% Tween 20 pH 7.5.

MABTB: MABT + 2% blocking powder. Make a 10% stock solution by heating 1 g of blocking powder (Boehringer, 1096176) in 10 ml of MABT. Dissolve, autoclave, and store in aliquots at –20 °C.

Magenta-phos™: dissolve at 50 mg/ml in dimethylformamide. Keep at –20 °C in the dark. Store at –20 °C or better at –70 °C. Magenta-phos is stable for at least a year at –70 °C.

Mouse embryo powder: homogenize ~ 12.5–14.5-day-old mouse embryos in a minimum volume of PBS. Add 4 vol. of ice-cold acetone, mix, and incubate on ice for 30 min. Spin at 10 000 *g* for 10 min and remove supernatant. Wash pellet with ice-cold acetone, spin again, and remove the supernatant. Spread the pellet out and grind it into a fine powder on a sheet of filter paper and allow it to air dry. Store in an airtight tube at 4 °C.

NBT (4-nitroblue tetrazolium chloride; Boehringer, 1 087 479): dissolve at 75 mg/ml in 70% dimethylformamide. Store in aliquots at –20 °C.

Table 1. *Continued*

NBT/BCIP staining solution: add 4.5 µl of 75 mg/ml NBT in 70% dimethylformamide and 3.5 µl of 50 mg/ml BCIP in dimethylformamide to 1 ml of NTMT buffer.

NTMT: 100 mM NaCl, 100 mM Tris–HCl pH 9.5, 50 mM $MgCl_2$, 0.1% Tween 20 (if necessary, add levamisole to 5 mM). Make from concentrated stock solutions on the day of use (the pH will decrease on storage, due to absorption of carbon dioxide).

PFA fix: paraformaldehyde is dissolved in PBS at 65 °C. If it does not readily dissolve add a drop or two of 1 M NaOH solution to pH 7.5. It should be cooled to 4 °C and used within two days.

PBS (phosphate-buffered saline): 130 mM NaCl; 7 mM $Na_2HPO_4.2H_2O$; 3 mM $NaH_2PO_4.2H_2O$. For 10 × PBS mix 75.97 g NaCl, 12.46 g $Na_2HPO_4.2H_2O$, 4.80 g $NaH_2PO_4.2H_2O$. Dissolve in less than 1 litre of distilled water; adjust to pH 7 and final volume of 1 litre, sterilize by autoclaving.

PBT: PBS, 0.1% Tween 20.

Rabbit peroxidase anti-peroxidase, PAP (Jackson ImmunoResearch Laboratories, 323–005–024).

Sheep anti-digoxigenin–alkaline phosphatase (AP) Fab fragments: 150 U/200 µl (Boehringer Mannheim, 1 093 274).

Sheep anti-digoxigenin–horse-radish peroxidase (POD) Fab fragments: 150 U lyophilized (Boehringer Mannheim, 1 207 733).

Sheep anti-fluorescein–alkaline phosphatase (AP) Fab fragments: 150 U/200 µl (Boehringer Mannheim, 1 426 338).

Sheep anti-fluorescein–horse-radish peroxidase (POD) Fab fragments: 150 U lyophilized (Boehringer Mannheim, 1 426 346).

sHybe: 50% formamide, 1.3 × SSC, 5 mM EDTA pH 8, 50 µg/ml yeast RNA, 0.2% Tween 20, 0.5% CHAPS, 100 µg/ml heparin, adjust to pH 6.5 with 1 M citric acid (modified from ref. 5).

Sigma *Fast*™ Fast Red (Sigma, F 4648): dissolve a buffer tablet in water, add a stain tablet, and use immediately.

TrueBlue™ peroxidase substrate staining solution (KPL, Kirkegaard & Perry Laboratories, 71–00–64): use as supplied. Store at 4 °C.

Tween 20 (Sigma, P 1379): 20% solution in sterile deionized water. Store at room temperature.

Vectastain ABC Elite Rabbit IgG kit: contains biotinylated goat anti-rabbit, avidin DH, biotinylated peroxidase (Vector Laboratories, PK-6101).

Vector™ Red alkaline phosphatase substrate (Vector Labs, SK-5100): mix stock solutions as described with kit.

Yeast RNA (Sigma, R 6750): dissolve in sterile deionized water at 50 mg/ml. Store in aliquots at –20 °C.

Zebrafish acetone powder: grind 4 g of adult fish in liquid nitrogen with a pre-cooled mortar and pestle. Add 16 ml of 0.85% saline and then 16 ml of cold (–20 °C) acetone. Mix vigorously and keep at 4 °C for 30 min. Collect the precipitate by centrifugation at 10 000 *g* for 10 min. Resuspend in fresh cold acetone and allow to sit at 4 °C for 10 min. Respin at 10 000 *g* for 10 min. Transfer to clean filter paper (Whatman Grade 50 Hard) on aluminium foil. Allow to dry, spreading and dispersing until dry. Grind again with mortar and pestle. Collect powder. Transfer to airtight container. The yield is 10–20% of the original wet weight.

zHybe: 50–65% formamide, 5 × SSC, 500 µg/ml yeast RNA, 0.1% Tween 20, 50 µg/ml heparin, adjust to pH 6.5 with 1 M citric acid.

for *Drosophila*, zHybe for zebrafish, and sHybe for mouse, chick, and *Xenopus*. Post-hybridization washes should be performed as described in Chapter 4. The protocols in this chapter all start with a blocking step to prevent non-specific binding of the antibody(s) used in the visualization of the signal.

1.1 Two colour *in situ* hybridization with chromogenic substrates for alkaline phosphatase and horse-radish peroxidase

Antisense RNA probes labelled with fluorescein and digoxigenin can be localized with antibodies conjugated with different enzymes. The most commonly used enzymes are horse-radish peroxidase and alkaline phosphatase. The signals are visualized with different chromogenic substrates which produce insoluble precipitates.

In general, fluorescein labelled probes are less stable than those labelled with digoxigenin and so the fluorescein labelled probe should be visualized first. It is best to hybridize both probes simultaneously and wash off unbound probes in the same manner as for single probes. It is possible to add a mixture of the two antibodies since they are conjugated with different enzymes. However, more consistent results and stronger signals are obtained when the antibody incubations and visualization steps are performed sequentially. In this case the fluorescein labelled probe should be visualized first.

If the antibody is conjugated with horse-radish peroxidase then the best substrates are diaminobenzidine (DAB) and TrueBlue™. The former gives a dark brown precipitate while the latter forms a bright blue precipitate. The DAB precipitate is insoluble in aqueous and organic solvents, whereas the TrueBlue™ precipitate is much less stable and specimens should be photographed immediately. However, the TrueBlue™ staining solution is considerably more sensitive than DAB, requiring that the antibody be used at a 50-fold greater dilution than for DAB. If DAB is the chosen chromogen, it is best to use it to visualize the fluorescein labelled probe. If TrueBlue™ is to be used, then it must be the last visualization step. This requires that it is used to identify the anti-digoxigenin Fab fragments conjugated to horse-radish peroxidase. The extra sensitivity of TrueBlue™ allows it to be used on larger embryos such as chick and mouse.

These horse-radish peroxidase substrates are used in combination with a substrate for alkaline phosphatase which produce a contrasting, coloured precipitate. DAB is best used in conjunction with a mixture of nitroblue tetrazolium and X-phosphate (BCIP, 5-bromo-4-chloro-3-indolyl-phosphate). The latter gives a dark blue/purple precipitate which is insoluble in aqueous and organic solutions. The Fast Red alkaline phosphate substrates produce a bright red precipitate which contrasts well with the TrueBlue™ precipitate.

Protocol 1.   Zebrafish or *Drosophila* two colour whole mount staining with DAB and BCIP/NBT[a]

Reagents

- 3% hydrogen peroxide: made fresh from a 30% stock solution stored at 4°C
- Anti-DIG–HRP: sheep anti-digoxigenin Fab fragments conjugated with horse-radish peroxidase, make a working dilution in blocking solution
- Anti-FLU–AP: sheep anti-fluorescein Fab fragments conjugated with alkaline phosphatase, make a working dilution in blocking solution

- Blocking solution for antibodies (see *Table 1*)
- DAB staining solution (see *Table 1*)
- NBT/BCIP staining solution (see *Table 1*)
- NTMT buffer (see *Table 1*)
- PBS (phosphate-buffered saline) (see *Table 1*)
- PBT: PBS, 0.1% Tween 20
- PFA fix (see *Table 1*)

Method

1. Carry out probe synthesis, tissue fixation and pre-treatment, hybridization, and washing as described in Chapter 4, *Protocols 1–3*. After *Protocol 3*, step 5 (in Chapter 4) replace the washing solution with blocking solution, and incubate at room temperature for at least 1 h on an orbital shaker.

2. Incubate for 2 h in pre-absorbed[b] anti-DIG–HRP in blocking solution at a dilution of 1:200 (0.75 U/ml).

3. Wash for 2 h in PBT (eight times for 15 min each).

4. Incubate for 2 min in DAB staining solution.

5. Add 1/1000 volume of 3% hydrogen peroxide to each incubation. Monitor the staining reaction and stop by rinsing thoroughly with PBT.

6. Replace the PBT with blocking solution and incubate at room temperature for up to 1 h on an orbital shaker.

7. Incubate for 2 h in anti-FLU–AP at a dilution of 1:2000 (0.375 U/ml) in blocking solution.

8. Wash for 2 h with PBT (eight times for 15 min each).

9. Wash three times for 5 min each in freshly made NTMT buffer.

10. Stain with NBT/BCIP staining solution.

11. Stop the reaction by washing with PBT.[c]

12. Fix the stain in PFA fix in PBS overnight.[d]

[a] This protocol is modified from refs 1 and 2.
[b] Pre-absorption is not necessary for *Drosophila* embryos or for zebrafish embryos which are less than 30 h old.
[c] The staining reaction can take from 10 min to several hours. If the staining reaction is to take several hours, it is convenient to perform all the antibody washes the day before staining and leave the embryos in PBT overnight. This allows a full working day to monitor the development of the stain.
[d] If the stain is not fixed, prolonged exposure to light can cause a dark background to develop.

Protocol 2. Chick, mouse, and *Xenopus* two colour whole mount staining with Fast Red and TrueBlue[a]

Equipment and reagents

- Orbital shaker or roller
- Anti-DIG–HRP: sheep anti-digoxigenin Fab fragments conjugated with horse-radish peroxidase, make a working dilution in blocking solution
- Anti-FLU–AP: sheep anti-fluorescein Fab fragments conjugated with alkaline phosphatase, make a working dilution in blocking solution

- Blocking solution: MABTB with 20% heat treated sheep serum (56°C for 30 min)
- MABT buffer (see *Table 1*)
- MABTB (see *Table 1*)
- Pre-stain buffer: 100 mM Tris–HCl pH 8.2
- TrueBlue™ staining solution (see *Table 1*)
- Vector Red™ staining solution (see *Table 1*)

Method

1. After hybridization and washing off the unbound probes as described in Chapter 4, *Protocol 4*, steps 1–4, replace the MABT with blocking solution (MABTB).[b]

2. Incubate for 1 h at room temperature.

3. Replace the MABTB with blocking solution and incubate for a further 1–2 h.

4. Incubate for at least 2 h in a 1:2000 to 1:8000 dilution (0.375–0.094 U/ml) of pre-absorbed anti-FLU–AP in blocking solution.

5. Rinse three times with MABT.

6. Wash three times for 60 min each with 10–20 ml MABT, by rolling at room temperature.

7. Wash twice for 10 min each with 100 mM Tris–HCl pH 8.2.

8. Incubate in Vector Red™ staining solution for a few hours to overnight.

9. When the colour has developed to the desired extent wash three times with PBT.

10. Replace PBT with blocking solution and incubate for a further 1 h.

11. Incubate for at least 2 h with pre-absorbed anti-DIG–HRP at a dilution of 1:2000 (0.075 U/ml) in blocking solution.

12. Rinse three times with MABT.

13. Wash three times for 60 min each with 10–20 ml MABT, by rolling at room temperature.

14. Replace the MABT with TrueBlue™ staining solution.[c]

15. Monitor staining and photograph. The TrueBlue™ colour may fade

once the staining solution is removed. It can be recovered by placing back in staining solution.

[a] Modified from refs 2–5.
[b] Block sample completely: intense endogenous peroxidase activity may be blocked by incubating slides or tissues for 30 min in 0.3% (w/v) H_2O_2 in 100% methanol, followed by a 10–15 min rinse in 0.1 M Tris–HCl pH 7.6.
[c] TrueBlue™ is 10–50 times more sensitive than DAB. Hence initially use a tenfold dilution of the peroxidase-conjugated antibody. If a high background is seen repeat with a reduced antibody titre. Also, the stain may fade if the embryos are removed from the staining solution or fixed. If this occurs re-equilibrate in PBT and restain.

1.2 Two colour *in situ* hybridization with chromogenic substrates for alkaline phosphatase

As an alternative to using antibodies which are conjugated with different enzymes, it is possible to sequentially incubate in antibodies conjugated with alkaline phosphatase. This requires that, after visualizing the signal from the fluorescein probe with the first antibody and substrate, the enzyme must be inactivated either by heating or by low pH treatment. It is important that the enzyme is completely inactivated otherwise, when the second substrate is added, it will precipitate over the first signal as well as over the location of the second probe. This might be interpreted as being caused by co-localization of the two target transcripts. Thus controls must be performed to confirm that overlapping signals are caused by co-localization of RNAs.

The most commonly used alkaline phosphatase substrate combinations are NBT/BCIP and a Fast Red. The NBT/BCIP substrate is the most sensitive and should be used for the weaker signal. There are several different Fast Red substrate kits commercially available (1, 6, 7). They all differ slightly in their sensitivity and the backgrounds that they produce (2).

(a) Vector™ Red (Vector Laboratories) is supplied as a kit containing three solutions which are stored at 4°C. The solutions must be mixed immediately before use. Staining with Vector Red is quite rapid, appearing within a few minutes to 1–2 h. Prolonged incubation tends not to intensify the signal. The yolk in zebrafish embryos stains yellow but the background in the embryo remains low. The red precipitate is stable to heat and so the alkaline phosphatase can be inactivated by heat treating at 65°C for 30 min.

(b) The Fast Red substrate from Boehringer Mannheim is supplied in the form of tablets which can be stored for prolonged periods at –20°C. The tablet is dissolved in Tris–HCl pH 8.2 buffer immediately prior to use. The solution throws down a precipitate after prolonged incubation. The signal develops less quickly than Vector™ Red but produces a more intense red precipitate. The background can be quite orange in both the yolk and the embryo. The red precipitate is heat labile and so the alkaline phosphatase must be inactivated by incubating in 100 mM glycine–HCl pH 2.2 for 30 min. The background staining can be reduced by including polyvinyl alcohols (PVA) in the staining reaction. This does, however, slightly reduce the intensity of the signal (2).

(c) Sigma *Fast*™ Fast Red is supplied as tablets for the Tris buffer and the substrate. It gives similar results to the product from Boehringer in that the signal is intense and develops slowly. Backgrounds are a little less than the latter but greater than with Vector™ Red. It also produces less precipitate in the staining solution. The precipitate is sensitive to heat and so the alkaline phosphatase must be inactivated by acidic glycine treatment.

If a combination of Fast Red and NBT/BCIP is to be used, the fluorescein labelled probe must be visualized first. The Fast Red substrate should be used first to identify the less abundant transcript. This means that, if the expression pattern of the second transcript is overlapping with the first, the second staining reaction can be carefully monitored and the blue/purple precipitate will appear over the red. This allows the reaction to be stopped before the blue/purple precipitate completely masks the red. If co-localized signals are expected, then two controls should be performed alongside the main experiment to ensure that all specimens receive the same treatments. One control should have the first antibody omitted and the second should have the second antibody omitted, but otherwise they should be treated exactly as those undergoing the complete procedure. Signal in the region of overlap in the first control and not the second should indicate that the staining is from the second probe and not from incompletely inactivated first antibody.

Protocol 3. Sequential alkaline phosphatase staining with chromogenic substrates of zebrafish embryos[a]

Reagents

- Anti-DIG–AP: sheep anti-digoxigenin Fab fragments conjugated with alkaline phosphatase, make a working dilution in blocking solution
- Anti-FLU–AP: sheep anti-fluorescein Fab fragments conjugated with alkaline phosphatase, make a working dilution in blocking solution
- AP inactivation solution (see *Table 1*)
- Blocking solution for antibodies (see *Table 1*)

- Fast Red (see *Table 1*)
- NBT/BCIP staining solution (see *Table 1*)
- NTMT buffer (see *Table 1*)
- PFA fix (see *Table 1*)
- Pre-stain buffer: 100 mM Tris–HCl pH 8.2, 0.1% Tween 20
- Sigma *Fast*™ Fast Red (see *Table 1*)
- Vector Red™ staining solution (see *Table 1*)
- Washing solution: 2 mg/ml BSA, 1% DMSO in PBT

Method

1. After hybridization and washing off unbound probes as described in Chapter 4, *Protocol 3*, steps 1–5, replace the PBT with blocking solution. Incubate at room temperature for 1 h.

2. Replace the blocking solution with a 1:5000 dilution (0.15 U/ml) of anti-FLU–AP in blocking solution. Incubate for 2 h.

3. Wash for 2 h in washing solution (eight times for 15 min each).

4. Equilibrate three times for 5 min each in pre-stain buffer (100 mM Tris–HCl pH 8.2, 0.1% Tween 20).

5. Stain with Vector™ Red, Fast Red, or Sigma *Fast*™ Fast Red.

6. Stop the reaction by washing in PBT.

7. Rinse in PBT, and heat to 65°C for 30 min to inactivate the alkaline phosphatase if stained with Vector™ Red. For Fast Red (Boehringer) or Sigma *Fast*™ Fast Red incubate in AP inactivation solution for 30 min, and then thoroughly wash in PBT.

8. Fix the stain in PFA fix for 20 min.

9. Block with blocking solution for 60 min.

10. Incubate for 2 h with a 1/5000 dilution (0.15 U/ml) of anti-DIG–AP in blocking solution.

11. Wash eight times for 15 min each in PBT to remove unbound antibody.

12. Equilibrate for three times for 5 min each in NTMT buffer.

13. Visualize the signal by incubating in NBT/BCIP staining solution.

14. Stop the reaction by rinsing in PBT and fixing with PFA fix for 20 min.

[a] Modified from that described in refs 1, 2, 9–11.

An alternative to Fast Red and NBT/BCIP is to use Magenta-phos™ (Biosynth AG) and BCIP (8). Magenta-phos™ produces a magenta coloured precipitate which is insoluble in aqueous and organic solutions. The magenta contrasts well with the light blue/turquoise precipitate generated by BCIP in the absence of NBT. If the intensity of each precipitate is similar then regions of overlap are a distinct shade of dark blue. Either substrate can be used first in sequential alkaline phosphatase staining. The coloured precipitates with both these substrates take a long time to develop and staining may take one to two days at 37°C. The staining times can be reduced by combining Magenta-phos™ with a 50-fold dilution of the Vector alkaline phosphatase substrate kit III (2).

Protocol 4. Sequential alkaline phosphatase staining with chromogenic substrates of chick, mouse, and *Xenopus* embryos[a]

Equipment and reagents

- Orbital shaker or roller
- AP inactivation solution (see *Table 1*)
- Blocking solution: MABTB with 20% heat treated sheep serum (56°C for 30 min)
- Anti-DIG–AP: sheep anti-digoxigenin Fab fragments conjugated with alkaline phosphatase, make a working dilution in blocking solution

Protocol 4. *Continued*

- Anti-FLU–AP: sheep anti-fluorescein Fab fragments conjugated with alkaline phosphatase, make a working dilution in blocking solution
- Fast Red (Boehringer, 1 496 549): dissolve a tablet in 2 ml of 100 mM Tris–HCl pH 8.2
- MABT buffer (see *Table 1*)
- MABTB (see *Table 1*)
- NBT/BCIP staining solution (see *Table 1*)
- NTMT buffer (see *Table 1*)
- PFA fix (see *Table 1*)
- Pre-stain buffer: 100 mM Tris–HCl pH 8.2, 0.1% Tween 20
- Sigma *Fast*™ Fast Red (see *Table 1*)
- Vector Red™ staining solution (see *Table 1*)

Method

1. After hybridization and washing off the unbound probes as described in Chapter 4, *Protocol 4*, steps 1–4, replace the MABT with MABTB.
2. Incubate for 1 h at room temperature with gentle shaking or rolling.
3. Replace with blocking solution and incubate for a further 1–2 h with gentle shaking or rolling.
4. Replace the blocking solution with a 1:5000 dilution of pre-absorbed anti-FLU–AP and incubate overnight at 4°C.
5. Rinse three times with MABT.
6. Wash three times for 60 min each with 10–20 ml MABT by rolling at room temperature.
7. Wash at least twice for 10 min each in 100 mM Tris–HCl pH 8.2.
8. Incubate with Fast Red, Sigma *Fast*™ Fast Red, or Vector™ Red staining solution. Rock for the first 20 min and then transfer to 24-well plates for observation.
9. When the colour has developed to the desired extent (1 h to overnight) wash three times with PBT.
10. Inactivate the alkaline phosphatase by incubating for 30 min in AP inactivation solution. Then incubate in PFA fix overnight at 4°C to preserve the stain.
11. Repeat the block by incubating in blocking solution for 1–2 h.
12. Replace with fresh solution containing a 1:2000 to 1:5000 dilution of pre-absorbed anti-DIG–AP and incubate overnight at 4°C or 4 h at room temperature.
13. Rinse briefly three times with MABT.
14. Wash three times for 60 min each with 10–20 ml MABT, by rolling at room temperature.
15. Wash at least twice for 10 min each in NTMT.
16. Incubate with 1.5 ml of NBT/BCIP staining solution.[b]
17. Rock for the first 20 min and then transfer to 24-well plates for observation. Development is faster at 37°C.

[a] This protocol is modified from refs 3–5, 8.
[b] Increasing the concentration of Tween 20 from 0.1% to 1% gives a darker blue/purple precipitate.

1.3 Double fluorescent *in situ* hybridization

Transcripts which have overlapping domains of expression can be a problem to visualize with chromogenic enzyme substrates since the heavier or darker precipitate can mask the lighter one. In an attempt to overcome this problem fluorochromes may be used to visualize the signals. If the fluorochromes are enzyme substrates then they can be substituted for the chromogenic substrates with only slight modification to the protocols. Enzyme substrates have the advantage that they amplify the signal so that a single enzyme molecule will produce many insoluble substrate molecules. The Fast Red precipitates fluoresce strongly when viewed by epifluorescence with a rhodamine filter set. This can be used in combination with the Enzyme Labeled Fluorescence (ELF™) alkaline phosphatase substrate supplied by Molecular Probes (Eugene). This substrate is initially non-fluorescent and colourless but is converted by the enzyme to a poorly soluble crystalline precipitate which fluoresces yellow/green with a DAPI filter set. The ELF™ signal is usually greater than that of Fast Red and so should be used to visualize the second antibody.

Protocol 5. Double fluorescent *in situ* hybridization of zebrafish or *Drosophila* embryos[a]

Equipment and reagents

- Microscope equipped for epifluorescence with a rhodamine and DAPI filter set
- 0.5% Triton X-100 in PBS
- Pre-stain buffer: 100 mM Tris–HCl pH 8.2, 0.1% Tween 20
- Anti-DIG–AP: sheep anti-digoxigenin Fab fragments conjugated with alkaline phosphatase, make a working dilution in blocking solution
- Anti-FLU–AP: sheep anti-fluorescein Fab fragments conjugated with alkaline phosphatase, make a working dilution in blocking solution
- AP inactivation solution (see *Table 1*)

- Blocking solution: 2 mg/ml BSA, 5% sheep serum, 1% DMSO, 0.1% Triton X-100 in PBS
- ELF™ pre-reaction buffer: 30 mM Tris–HCl pH 7.5, 150 mM NaCl
- ELF™ stop buffer (see *Table 1*)
- ELF™–AP substrate kit (Molecular Probes, E-6601): includes ELF™ reagent, reaction buffer, aqueous mounting medium
- Fast Red (see *Table 1*)
- PFA fix (see *Table 1*)
- Sigma *Fast*™ Fast Red (see *Table 1*)
- Vector Red™ staining solution (see *Table 1*)

Method

1. After hybridization and washing off the unbound probes as described in Chapter 4, *Protocol 3*, steps 1–5, replace the wash with blocking solution.

2. Incubate for at least 60 min.

3. Incubate in a 1:2000 dilution (0.37 U/ml) of anti-FLU–AP in blocking solution overnight at 4°C.

4. Wash embryos with PBT for 2 h (eight times for 15 min each).

117

Protocol 5. *Continued*

5. Equilibrate with pre-stain buffer at room temperature by washing three times for 5 min each.

6. Stain embryos with Fast Red (Boehringer), Sigma *Fast*™ Fast Red, or Vector™ Red.

7. Stop the reaction by washing several times with PBT.

8. Inactivate the alkaline phosphatase activity by incubating in AP in-activation solution twice for 15 min each at room temperature.

9. Rinse four times for 5 min each with PBT.

10. Incubate in PFA fix for 20 min at room temperature.

11. Rinse five times for 5 min each with PBT.

12. Incubate embryos in blocking solution for 60 min.

13. Incubate in a 1:5000 dilution (0.15 U/ml) of anti-DIG–AP in blocking solution overnight at 4°C.

14. Wash in 0.5% Triton X-100 in PBS for 2 h.[b]

15. Wash three times for 5 min each at room temperature with the ELF™ pre-reaction buffer.

16. For staining incubate in a 1:20 dilution of the ELF™ substrate reagent at room temperature for 5 h, or a 1:100 dilution of the ELF™ substrate overnight at room temperature.[c]

17. Monitor the staining reaction using a UV fluorescent microscope with a DAPI filter set at intervals after starting the reaction.

18. Stop the reaction by washing with ELF™ stop buffer.

19. Mount the tissue with the special aqueous mounting medium supplied with the kit from Molecular Probes.[d]

[a] Procedure modified from refs 2, 9–11.
[b] DMSO and Tween 20 in the wash solutions cause the final ELF crystals to be large.
[c] This is far longer than recommended by Molecular Probes. With zebrafish embryos it is not necessary to add 1 mM levamisole.
[d] This mountant preserves the ELF™ signal better than the usual glycerol-based aqueous mountants for fluorochromes.

With *Drosophila* embryos it is possible to directly visualize a hybridized antisense RNA probe which is labelled with fluorescein. The signals are weak as there is no enzymatic amplification of the signal. The backgrounds with fluorescein can also be quite high, relative to the signal, but this can be improved by image capture and enhancement or by using the confocal micro-scope. The non-confocal method requires a compound microscope equipped for epifluorescence with fluorescein and rhodamine filter sets and fitted with a motor-driven automatic focus. The latter is controlled by Biovision DCi

(Digital Confocal Imaging) software (Improvision, Coventry) which also captures images via a cooled CCD camera. The software captures images at different focal planes. These are processed by the software and recombined to produce a 3D image. This method has proven successful with *Drosophila* embryos viewed by epifluorescence but with larger embryos the background fluorescence is so great that a confocal microscope must be used.

The method in *Protocol 6* is based on original procedures from Diethard Tautz, Bruce Edgar, Philip Ingham, and David Ish-Horowicz with modifications by Sheena Pinchin. A combination of fluorescein and digoxigenin labelled probes is hybridized to the embryos. There is no need to treat the embryos with proteinase K after rehydration but it is important to refix them to minimize the loss of target RNA. Fluorescein labelled RNA is unstable in light and also at low pH, therefore the tubes are kept in the dark as much as possible during the treatment. The post-hybridization washes are kept as short as possible and the hybridization buffer has a pH of 6.5. The fluorescein labelled probe is viewed directly by epifluorescence and the digoxigenin labelled probe is localized with anti-digoxigenin antibody conjugated with alkaline phosphatase and stained weakly with Vector™ Red.

Protocol 6. Direct fluorescence *in situ* hybridization to *Drosophila* embryos

Equipment and reagents

- Orbital shaker or roller
- Confocal microscope
- 100 μm and 670 μM nylon mesh sieves
- 4% formaldehyde in 1 × PBS: made from a 40% stock solution of formaldehyde and 10 × PBS
- 6% sodium hypochlorite solution
- Anti-DIG–AP: anti-digoxigenin Fab fragments conjugated with alkaline phosphatase, make a working dilution in blocking solution

- Blocking solution: PBT containing 0.2 mg/ml BSA and 0.025% sodium azide
- Citifluor AF3 (Agar Scientific)
- dHybe: 50% formamide, 5 × SSC, 100 μg/ml yeast RNA, 0.1% Tween 20, 50 μg/ml heparin, adjust to pH 6.5 with 1 M citric acid
- Heptane
- Methanol chilled on dry ice
- Pre-stain buffer: 100 mM Tris–HCl pH 8.2
- Vector Red™ staining solution (see *Table 1*)

A. *Preparation of Drosophila embryos*

1. Collect *Drosophila* embryos on yeasted agar plates in a population cage and age as required.

2. Wash off embryos with water and a soft paintbrush. Pour through a coarse (670 μm) nylon sieve to remove any dead flies. Collect the embryos in a fine (100 μm) nylon sieve.

3. Wash embryos from the sieve into a sterile 50 ml screw-capped, polypropylene centrifuge tube with 6% sodium hypochlorite solution. Dechorionate for about 2 min (the exact time will depend on the age and strength of the bleach solution).

Protocol 6. *Continued*

4. Collect the dechorionated embryos in a fine sieve and rinse well with water to remove all traces of the bleach.

5. Transfer the embryos to a sterile 50 ml screw-capped, polypropylene centrifuge tube with 4% formaldehyde in PBS. Add an equal volume of heptane and mix on a roller for 30 min at room temperature.

6. Stop rolling and allow the embryos to settle at the interface. Remove most of both solutions with a pipette and replace with fresh fix and heptane. Incubate for a further 30 min on the roller. (Alternatively fix for 30 min with equal volumes of heptane and 10% formaldehyde in PBS.)

7. Remove the lower phase and replace with an equal volume of methanol at −20°C.[b] The embryos should pop out of their vitelline membranes within 1–2 min. They then fall to the bottom of the bottle.

8. Remove most of the methanol and replace with fresh cold methanol. Discard any embryos which have not sunk to the bottom of the tube.

9. Store the embryos in methanol at −20°C.

B. *Pre-hybridization treatments of Drosophila embryos*

1. Rehydrate through a graded methanol/PBS series:
    (a) 2 min in methanol:PBS, 7:3.
    (b) 2 min in methanol:PBS, 5:5.
    (c) 2 min in methanol:PBS, 3:7.
    (d) 2 min in PBS.

2. Refix for a further 20 min in 4% formaldehyde in PBS.

3. Rinse with PBS. The embryos are now ready for hybridization without proteinase K treatment if hybridized at 70°C.

C. *Hybridization of Drosophila embryos*

1. Rinse five times in PBT.

2. Rinse in 1:1 PBT:dHybe.

3. Pre-hybridize in dHybe for at least 1 h at 70°C.

4. Hybridize overnight at the appropriate temperature in 150–400 µl of dHybe in 1.5 ml microcentrifuge tubes containing heat denatured probe. For double label *in situ* hybridizations mix digoxigenin and fluorescein probes. Use fluorescein antisense probes at a 1:50 dilution of a standard reaction that has been precipitated and redissolved in 100 µl of DEPC water. (Note this is fourfold higher than recommended for detection of hybridized signals with antibodies.)

D. *Post-hybridization washes of Drosophila embryos*

1. Rinse in dHybe at 70°C.

2. Wash for 20 min in dHybe at 70°C.
3. Wash in 1:1 PBT:dHybe for 20 min at 70°C.
4. Wash in PBT four times for 20 min each at room temperature.
5. Mount a few embryos in 0.1 ml of Citifluor, spacing the coverslip and slide with tape, and sealing with nail polish. Examine under the microscope using epifluorescence with a fluorescein filter set.

E. *Visualization of the digoxigenin labelled probe*
  1. Block non-specific binding sites by washing in blocking solution for 20 min.
  2. Incubate in a 1:2000 dilution of anti-DIG–AP for 3 h at room temperature on a gently rocking platform.
  3. Rinse twice with PBT.
  4. Wash five times with PBT over a period of 3 h, on a gently rocking platform.
  5. Rinse twice with pre-stain buffer.
  6. Transfer embryos to a 24-well microtitre plate, for easy observation of the staining reaction under a dissecting microscope.
  7. Stain samples with 0.4 ml Vector™ Red staining solution for 10–30 min.
  8. Stain until the colour is visible under the dissecting microscope, but not too intense. If the precipitate is too dense, it will quench the fluorescence from the fluorescein.
  9. Stop the staining reaction by washing twice in PBT.
  10. Mount in Citifluor.
  11. View by epifluorescence microscopy. The Vector™ red precipitate fluoresces red with a rhodamine filter set giving a high signal-to-noise ratio. The fluorescein signal is seen with a fluorescein filter set but is very much weaker with a higher background. Higher resolution is obtained with the confocal microscope.

[a] Modified from ref. 2.
[b] Better morphology is retained if the methanol is cooled to –70°C on dry ice.

2. Methods for simultaneous *in situ* localization of transcripts and tissue antigens

In addition to relating a specific mRNA to other gene transcripts it may be desirable to compare its location with its own translation product or that of another gene or tissue antigen. This requires the procedure of *in situ* hybridization to be followed by immunolocalization of an antigen. It is best, but not absolutely necessary, to perform the *in situ* hybridization first as this will reduce the likelihood that the target transcript will be lost by degradation with RNase. The immunlocalization requires an antibody to the particular tissue antigen which will recognize it after the tissue has been through the rigours of *in situ* hybridization. For this reason, each antibody must be tested independently as to its effectiveness to identify the antigen after the *in situ* procedure. In some cases the fixation of the material may have to be modified and the temperature of hybridization or the duration of hybridization shortened. The tissue antigen may also be susceptible to protease digestion so it may be advisable to substitute an acetone permeabilization step for proteinase K treatment (2). It is important to realize that not all tissue antigens or antibodies may be suitable for this combination of procedures.

The following protocols describe three different methods of immuno-localization of a tissue antigen. These should be used following a standard *in situ* hybridization with an antisense RNA probe labelled with digoxigenin and visualized with an antibody conjugated with alkaline phosphatase and stained with NBT/BCIP. In all cases the tissue antigen is stained with DAB, a horse-radish peroxidase substrate.

2.1 Indirect immunolocalization of a tissue antigen

This is the simplest of the three methods described and is the least sensitive, but, in so being, is also the least likely to lead to background problems. Following the *in situ* hybridization the tissue is blocked to prevent non-specific binding of the primary antibody and then incubated with a monoclonal antibody raised against the tissue antigen. The unbound antibody is washed off and a secondary antibody (goat anti-mouse IgG) added which is conjugated with horse-radish peroxidase. After washing off unbound antibody the material is stained with DAB. The blocking solution should ideally contain serum from the same animal in which the secondary antibody was raised.

Protocol 7. Immunolocalization with a horse-radish peroxidase-conjugated secondary antibody

Reagents

- 3% hydrogen peroxide: made fresh from a 30% stock solution stored at 4°C
- Anti-mouse IgG–HRP: goat anti-mouse IgG conjugated with horse-radish peroxidase
- Blocking solution for antibodies (see *Table 1*)
- DAB staining solution (see *Table 1*)
- PFA fix (see *Table 1*)
- Primary IgG monoclonal antibody

Method

1. Perform the *in situ* hybridization as described in Chapter 4. Use an antisense RNA probe labelled with digoxigenin and an antibody conjugated with alkaline phosphatase. Stain with NBT/BCIP as described in *Protocols 1, 3,* and *4*.

---

2. Stop the staining by rinsing in PBT and refix for 20 min in PFA fix.

3. Incubate in blocking solution for 30 min at room temperature.

4. Incubate for 5 h at room temperature in a 1:2000 dilution of primary monoclonal antibody.[a]

5. Wash for 2 h with PBT (eight times for 15 min each).

6. Incubate overnight at 4°C in blocking solution containing a 1:5000 dilution of anti-mouse IgG–HRP.

7. Wash for 2 h with PBT (eight times for 15 min each).

8. Incubate for 2 min in DAB staining solution.[b]

9. Add 1/1000 volume of 3% hydrogen peroxide.

10. Monitor the staining reaction and stop by rinsing thoroughly with PBT.

11. Fix the stain by incubating in PFA fix for 20 min.

---

[a] This is modified from a method described in ref. 2.
[b] The dilution of primary antibody used is dependent on the particular antibody and may have to be optimized for the highest signal-to-noise ratio.
[c] TrueBlue™ can be used as an alternative substrate to DAB. The TrueBlue™ horse-radish peroxidase substrate is more sensitive than DAB, so decrease the titre of the secondary antibody by 10- to 50-fold. However, the blue precipitate formed is less stable, being partially soluble in alcohol and water. If used in a two colour *in situ* hybridization with horse-radish peroxidase- and alkaline phosphatase-conjugated antibodies, the phosphatase should be stained before the peroxidase, otherwise the blue precipitate will be lost or weakened.

2.2 Immunolocalization by the peroxidase anti-peroxidase (PAP) method

This method uses a soluble peroxidase anti-peroxidase complex (PAP) first described by Sternberger *et al.* (12). The PAP complexes are formed from three peroxidase molecules and two anti-peroxidase antibodies and are used as a third layer in the visualization step. After the unbound primary antibody is washed off a second 'bridging' antibody is added in excess so one of its two available binding sites binds to the primary antibody and the other to the PAP complex. The immunoglobulin in the PAP complex must be raised in the same animal as that of the primary antibody. This method is 100–1000 times more sensitive than the indirect method in *Protocol 7*.

Protocol 8. Immunolocalization by the peroxidase anti-peroxidase (PAP) method[a]

Reagents

- 3% hydrogen peroxide: made fresh from a 30% stock solution stored at 4°C
- Blocking solution for antibodies (see *Table 1*)

Protocol 8. *Continued*

- DAB staining solution (see *Table 1*)
- PAP complex: peroxidase-conjugated rabbit anti-peroxidase (PAP, Jackson Immuno-Research Laboratories)
- PFA fix (see *Table 1*)
- Primary antibody raised in rabbit against the tissue antigen
- Secondary antibody: goat anti-rabbit IgG 'bridging antibody' (Jackson ImmunoResearch Laboratories)

Method

1. Perform the *in situ* hybridization as described in Chapter 4. Use an antisense RNA probe labelled with digoxigenin and an antibody conjugated with alkaline phosphatase. Stain with NBT/BCIP as in *Protocols 1, 3,* and *4.*

2. Stop the staining by rinsing in PBT and refix for 20 min in PFA fix in PBS.

3. Incubate in blocking solution for 30 min at room temperature.

4. Incubate 5 h at room temperature in a 1:2000 dilution of primary rabbit antibody in blocking solution.

5. Wash for 2 h with PBT (eight times for 15 min each).

6. Incubate overnight at 4°C in blocking solution containing a 1:100 dilution of secondary antibody in blocking solution.[b]

7. Wash for 2 h with PBT (eight times for 15 min each).

8. Incubate for 5 h in blocking solution containing a 1:400 dilution of PAP complex.

9. Wash for 2 h with PBT (eight times for 15 min each).

10. Incubate for 2 min in DAB staining solution.[c]

11. Add 1/1000 volume of 3% hydrogen peroxide.

12. Monitor staining and stop by rinsing thoroughly with PBT.

[a] This is modified from a method described in ref. 2.
[b] It is important in the PAP technique for the bridging antibody to be applied in excess. This way one arm of the divalent Fab portion of the immunoglobulin molecule can bind the primary antibody, while the other arm is free to bind the PAP complex.
[c] TrueBlue™ can be used as an alternative substrate to DAB. The TrueBlue™ horse-radish peroxidase substrate is more sensitive than DAB and so the titre of the antibody can be decreased by 10- to 50-fold. However, the blue precipitate formed is less stable, being partially soluble in alcohol and water. If used in a two colour *in situ* hybridization with horse-radish peroxidase- and alkaline phosphatase-conjugated antibodies, the phosphatase should be stained first and then the peroxidase. Otherwise the blue precipitate will be lost or weakened.

2.3 Immunolocalization by the ABC method

In this procedure incubation with a primary antibody is followed by a biotinylated secondary antibody and then a preformed avidin–biotinylated enzyme complex (ABC). Avidin has a very high affinity for biotin and binding of avidin to biotin is effectively irreversible. Avidin has four biotin binding sites which allows macromolecular complexes to form with biotinylated enzymes or antibodies. The following procedure utilizes a primary antibody raised in rabbit. The appropriate rabbit IgG Vector Elite™ kit should be chosen. It is also important to use the biotin-free serum supplied with the kit. Other sera may contain biotin which would lead to serious background problems.

Protocol 9. Avidin biotinylated enzyme complex staining of zebrafish embryos[a]

Reagents

- 3% hydrogen peroxide: made fresh from a 30% stock solution stored at 4°C
- Benzylbenzoate:benzyl alcohol mixed in the ratio 2:1
- DAB stock solution: 5 mg/ml diaminobenzidine in 10 mM Tris–HCl pH 7
- Primary antibody raised in rabbit against the tissue antigen

- Blocking solution: 5% goat serum (supplied with the Vector Elite™ rabbit IgG kit), 1% DMSO, 0.1% Tween 20 in PBS
- Vector Elite™ rabbit IgG kit, contains reagents A and B for forming the avidin–biotinylated peroxidase complex, biotinylated goat anti-rabbit, and goat serum (tested biotin-free)

Method

1. Perform the *in situ* hybridization as described in Chapter 4. Use an antisense RNA probe labelled with digoxigenin and an antibody conjugated with alkaline phosphatase. Stain with NBT/BCIP as in *Protocols 1, 3*, and *4*.

2. Stop the staining by rinsing in PBT and refix for 20 min in PFA fix.

3. Incubate in blocking solution for 30 min at room temperature.

4. Incubate for 5 h at room temperature in a 1/2000 dilution of primary rabbit antibody in blocking solution.

5. Wash four times for 20 min each with blocking solution.

6. Incubate in a 1:2000 dilution of secondary antibody (biotinylated goat anti-rabbit, Vector Elite™ rabbit IgG kit) for 4–5 h at room temperature or 4°C overnight.

7. Wash four times for 20 min each with blocking solution.

8. Rinse briefly with PBT.

9. Make the AB complex. Add 40 μl of reagent A (avidin DH) to 5 ml of blocking solution. Mix and add 40 μl of reagent B (biotinylated enzyme). Mix and incubate for 30 min at room temperature.

10. Incubate the embryos in the AB complex for 45–60 min.

11. Wash four times for 25 min each with blocking solution.

12. Wash briefly in PBT.

13. Incubate for 2 min in 2 ml PBT plus 100 μl of DAB stock solution.

Protocol 9. *Continued*

14. Add 4 μl of 3% H_2O_2. Allow to stain until the signal appears (usually a few minutes). Stop the reaction by washing the specimen in PBS.

15. Dehydrate by putting the specimen in 100% methanol. Change once (twice for 10 min each is enough). Clear the embryos by transferring to a 2:1 mixture of benzylbenzoate:benzyl alcohol.

ª This is modified from a method described in ref. 2.

References

1. Jowett, T. and Lettice, L. (1994). *Trends Genet.*, **10**, 73.
2. Jowett, T. (1996). *Tissue in situ hybridization: methods in animal development.* Wiley and Sons, NY.
3. Harland, R.M. (1991). *Methods Cell Biol.*, **36**, 685.
4. Lamb, T.M., Knecht, A.K., Smith, W.C., Stachel, S.E., Economides, A.N., Stahl, N., *et al.* (1993). *Science*, **262**, 713.
5. Henrique, D., Adam, J., Myat, A., Chitnis, A., Lewis, J., and Ish-Horowicz, D. (1995). *Nature*, **375**, 787.
6. Strähle, U., Blader, P., Adam, J., and Ingham, P.W. (1994). *Trends Genet.*, **10**, 75.
7. Hauptmann, G. and Gerster, T. (1994). *Trends Genet.*, **10**, 266.
8. Knecht, A.K., Good, P.J., Dawid, I.B., and Harland, R.M. (1995). *Development*, **121**, 1927.
9. Jowett, T. and Yan, Y-L. (1996). *Trends Genet.*, **12**, 387.
10. Jowett, T., Mancera, M., Amores, A., and Yan, Y-L. (1966). In *In situ hybridization: laboratory companion* (ed. M. Clark), p. 91. Chapman and Hall.
11. Jowett, T. and Yan, Y-L. (1966). In *A laboratory guide to RNA: isolation, analysis, and synthesis* (ed. P.A. Krieg), p. 381. Wiley–Liss, NY.
12. Sternberger, L.A., Hardy, P.H., Cuculis, J.J., and Meyer, H.G. (1970). *J. Histochem. Cytochem.*, **18**, 315.

6

Electron microscope *in situ* hybridization: the non-isotopic post-embedding procedure

FRANCINE PUVION-DUTILLEUL

1. Introduction

Although the advent of *in situ* hybridization for light microscopy goes back almost three decades (1, 2) and its extention to the electron microscope (EM) level occurred shortly after, it is only recently that this technique revolutionized our knowledge about the spatial distribution of defined nucleic acid molecules in the cell. Binder *et al.* (3) broke new ground in this field by concomitantly using for the first time hydrosoluble resin-embedded material, non-isotopic probes, and immunogold labelling of hybrids. This post-embedding non-radioactive procedure was immediately adopted, and became widely used after some procedural improvements which rendered it a method applicable to a wide variety of biological systems.

The EM *in situ* hybridization technique is a powerful tool for localizing specific nucleotide sequences among innumerable nucleic acid molecules and for identifying with certainty the structures in which the sequences under study are inserted so long as the methodology used is judicious. EM *in situ* hybridization can be applied to cryosections, prior to or after the embedding of cells, and on spread molecules. Each method has specific limitations. *In situ* hybridization on spread molecules is used rarely (4, 5) and due to technical difficulties is far from being a routine technique. We can expect that in the future this methodology will open doors to the molecular study of gene expression. *In situ* hybridization on ultrathin frozen sections (6) induces the unexpected and uncontrollable extraction of small molecules during the successive steps of the reaction which result in a rather distorted ultrastructure. Pre-embedding *in situ* hybridization (7) has crucial limitations which consist of:

(a) Uncontrollable extraction and artefactual rearrangements of molecules which occur both during the permeabilizing pre-treatment of samples and during the *in situ* hybridization treatment.

(b) Poor accessibility of targets due to restriction of diffusion of the probe.

(c) Bad preservation of ultrastruture rendering labelled areas unidentifiable.

The only limitation of the post-embedding *in situ* hybridization technique (8, 9) is that only the targets which are exposed at the surface of the ultrathin section bind the probe. Nevertheless, post-embedding *in situ* hybridization preserves the fine structure well and permits the clear identification of the structures containing the target under study, an advantage which warrants the use of *in situ* hybridization at the EM versus the optical level.

In this chapter we present a collection of procedures useful for the specific detection at the surface of ultrathin sections of embedded biological material of different kinds of nucleic acid molecules: cellular or viral molecules, DNA or RNA, double-stranded or single-stranded molecules, which have increased our knowledge of the functional compartmentalization of the cell nucleus. Only those procedures which have given satisfactory data in our hands and are routinely used in our laboratory are described.

2. Preparation of biological material for ultrathin sectioning

Cultured cells and small pieces of tissues are fixed, dehydrated, and embedded prior to being sectioned and processed for *in situ* hybridization (see *Table 1*).

2.1 Fixation

We found that glutaraldehyde fixation of cells allows the detection of single-stranded sequences including single-stranded portions of DNA molecules, RNA molecules of cellular or viral origin, and poly(A) tails. Formaldehyde fixation of cells permits additional detection of double-stranded DNA molecules, either cellular or viral. We carry out fixation as follows:

(a) Fix biological material (pellet of living cells, cell monolayer, piece of tissue) for 1 h at 4°C with either 1.6% glutaraldehyde (Taab Lab. Equip)

Table 1. Preparation of biological material for detection of RNA or DNA sequences by post-embedding EM *in situ* hybridization

Fixation	Formaldehyde (4%, 1 h) for DNA target
	Formaldehyde (4%, 1 h) or glutaraldehyde (1.6%, 1 h) for RNA
Dehydration	Methanol
Embedding	Lowicryl K4M
Polymerization	−30°C
Sectioning	Diamond knife
Support	Formvar carbon coated gold grids

---

or 4% formaldehyde (Merck) in 0.1 M Sörensen's phosphate buffer pH 7.2–7.4. If attached cells are used during the fixation step, detach the cells from their culture support by scraping with a piece of silicone and centrifuge the resulting cell suspension for 15 min at about 5000 *g* in a refrigerated centrifuge.

(b) At the end of the 1 h fixation step, rinse pellets and pieces of tissues in ice-cold phosphate buffer using three changes during a 2 h total period.

2.2 Dehydration and Lowicryl embedding

Methanol is preferred to ethanol and acetone for dehydrating the pellets of cells and pieces of tissues because no artefactual aggregation of viral DNA occurs with methanol (10). Lowicryl K4M (11) has been employed routinely in our laboratory since 1984 for embedding the cells so that we have accumulated a wide variety of eclectic embedded biological models which allow profitable retrospective studies on old blocks and comparison of data with those obtained with more recently embedded material. Other water miscible resin mixtures are available including Bioacryl described by Scala *et al.* (12), available commercially under the trademark name Unicryl, which also preserve fine structure, retain antigenicity (13), and provide good hybridization signals (14). The protocol used for Lowicryl K4M embedding is described in *Protocol 1* (15).

Protocol 1. Successive steps for Lowicryl K4M embedding

Equipment and reagents
- Fluorescence tubes (6 W, 360 nm UV light)
- Deep-freeze
- Lowicryl K4M (Polysciences Europe)
- Methanol
- Gelatin capsules

Method

1. Prepare the Lowicryl embedding solution by mixing 4 g cross-linker, 26 g monomer, and 0.15 g initiator. This solution is stored at –20°C and used within the same day.

2. Dehydrate the previously fixed and rinsed samples successively with 30% and 50% methanol, each for 5 min at 4°C.

3. Dehydrate and impregnate with Lowicryl K4M at –20°C. Use successively 70% methanol for 5 min, 90% methanol for 30 min, 1:1 (v/v) 90% methanol: Lowicryl K4M for 1 h, followed by 1:2 (v/v) 90% methanol: Lowicryl K4M for 1 h.

4. Plunge samples in Lowicryl K4M, first for 1 h, then in a second Lowicryl-bath overnight, each at –20°C.

Protocol 1. *Continued*

5. Embed in Lowicryl K4M using gelatin capsules, and store overnight at about −20°C.

6. Polymerize under long wavelength UV light for five days in a deep-freeze at −30°C and then for 24 h at room temperature by the use of Philips fluorescence tubes arranged in a sandwich fashion at 30 cm distance from the capsules.

2.3 Supports for ultrathin sections

Grids of gold or nickel are required for *in situ* hybridization. Copper reacts with the components of the hybridization solution, and therefore copper grids never can be used. Some investigators collect ultrathin sections directly on naked grids, often mesh 600 (16). We prefer employing 200 mesh gold grids coated with a Formvar film strengthened by carbon because of:

(a) The wider observation field.

(b) The high stability of the ultrathin sections under the beam even after additional extraction treatments.

(c) The good conservation of the grids bearing ultrathin sections which can be re-examined even after one or two years.

Formvar carbon coated grids are prepared as described in *Protocol 2*.

Protocol 2. Preparation of Formvar carbon coated grids

Equipments and reagents

- Glass slides and clear glass container
- Gold grids, 200 mesh
- Carbon evaporator
- Formvar
- 1,2-dichloroethane

Method

1. Prepare 80 ml of 3% (w/v) Formvar in water-free 1,2-dichloroethane.

2. To prepare the film, dip a glass slide into the solution, remove it slowly, and dry in a vertical position.

3. Float the film off the slide on to a water surface after scraping the edges of the slide, arrange the grids on the floating film, recover the film plus grids onto a piece of filter paper, and finally, pick up the paper carrying the film and the grids with forceps.

4. After air drying, evaporate a thin layer of carbon on the Formvar coated grids in order to stabilize the plastic film. The coated grids can be stored up to several months in a Petri dish.

2.4 Sectioning and recovering ultrathin sections

Ultrathin sections are made with an ultramicrotome and a diamond knife. They are collected on the Formvar carbon coated grids. Grids with ultrathin sections are placed in a box which is stored in darkness at room temperature until use for *in situ* hybridization (up to two years). As mentioned above (see Section 2.3) ultrathin sections also can be recovered onto naked gold or nickel grids but such samples are very weak and often become altered and torn during the successive steps of the *in situ* hybridization procedure.

3. Probes

3.1 Generalities

In situ hybridization is based on the pairing reaction between the nucleic acid target sequences which are under study in a biological sample and the complementary labelled DNA or RNA molecules (probes) of a hybridization solution. The procedures of detection and quantitation of specific target sequences by *in situ* hybridization are well established.

Radioactive probes, which have been used successfully for a long time, have four main limits:

(a) They are inherently unstable.

(b) The technique of high-resolution autoradiography is difficult.

(c) It requires exposure of autoradiograms for several months.

(d) Provides only low resolution.

Therefore, radioactive probes are now supplanted by non-isotopic probes and subsequent gold labelling of the hybrids.

The non-radioactive *in situ* hybridization procedure, which is simpler, more rapid, and gives a better resolution than autoradiography, represents a marked advance in EM *in situ* hybridization since it permits precise localization with respect to small substructures such as ribosomes and viruses. Since several haptens and different sizes of gold particles are available to label the probes and to detect hybrids, respectively, multiple detections can be performed on the same ultrathin section. In addition, *in situ* hybridization and immunogold detection of antigens can be performed concomitantly. All these advantages and possibilities render EM non-isotopic *in situ* hybridization very informative for revealing the composition of a defined structure.

3.2 Probe choice

3.2.1 Choice of the non-radioactive marker

Several reporter molecules, which include biotin (17), digoxigenin (18–20), bromodeoxyuridine (21, 22), or horse-radish peroxidase (23), can be incorporated into a nucleic acid probe by different approaches: enzymatic, chemical, or photochemical. Biotin is the most widely used hapten. Biotinylated probes are stable for several years (more than eight years) at $-20\,°C$. Indeed, adenovirus and herpes simplex virus DNA probes biotinylated by nick translation and purchased from Enzo (Enzo Biochem. Inc.) have been stored in our laboratory since 1988 without diminution of their activities. It must be emphasized, however, that the use of biotinylated probes is imperatively restricted to biological samples devoid of detectable endogenous biotin. In our hands, biotinylated probes specifically reveal defined RNA or DNA sequence targets when applied on Lowicryl K4M sections of syncytial plasmodium of Physarum polycephalum (24, 25), rabbit fibroblast cell line (26–28), mouse 3T3 cell line (29, 30), human HeLa cell line (31–34), human Hep2 cell line, and human skin (in preparation). Some tissues known to contain biotin such as liver and kidney (35) require the use of other non-isotopic probes.

3.2.2 Choice of the probe type

Both DNA and RNA probes are suitable to reveal specific RNA and DNA sequences by *in situ* hybridization. We extensively use double-stranded DNA probes which are biotinylated by nick translation (see Chapter 7, *Protocol 2*). Because of the resulting labelling of the two DNA strands, an intense hybridization signal is obtained over specific double-stranded DNA sequence targets such as ribosomal genes (25, 36, 37), viral genomes (26, 28, 38), and human *Alu* sequences (31, 32). Such probes, which in contrast to RNA probes are not strand-specific, also allow the detection of RNA molecules originating from the two DNA strands, a phenomenon which occurs in viral infection due to the symmetrical transcription of the viral genome. For example, with the same adenovirus or herpes simplex virus DNA probe, it is possible to detect specifically viral genomes (*Figure 1*), or their viral single-stranded (*Figure 2a*) or double-stranded portions (*Figure 2b*), or viral RNA (*Figures 3* and *4*). Each double-stranded DNA probe must be denatured before use, generally through a heat treatment which does not destroy haptens.

Synthetic oligonucleotide probes labelled by the 3' end tailing procedure with terminal deoxynucleotidyl transferase (see Chapter 2) are fruitful to detect a short, defined portion within a RNA molecule provided that the target sequence is known to permit the preparation of the probe, and is accessible to the probe (i.e. not located within secondary structure which would prevent hybrid formation). Since the oligonucleotide probes are single-stranded, a heat treatment before use is not required. A similarly labelled poly(dT) probe has proven to be useful for the localization of poly(A) tails of cellular and viral RNA (33, 34, 39–42).

Some investigators employ single-stranded RNA probes (see Chapters 3 and 4) to detect RNA sequences, either messenger or ribosomal RNA sequences (43), and DNA sequences. Depending on the direction of *in vitro* transcription either an antisense or a sense RNA probe is obtained. The anti-sense probe reveals the RNA molecules whereas sense probe is used as a negative control of hybridization. Both probes form RNA:DNA hybrids when *in situ* hybridization is performed after denaturation of a double-stranded DNA target. An advantage of single-stranded antisense and sense RNA probes is that they do not require denaturation before use. Nevertheless a heat treatment of riboprobes has proven to be useful to remove secondary structures resulting from intramolecular base pairing. Unfortunately, RNA probes non-specifically bind to supports including resin so that a significant background occurs. This can be reduced by a post-hybridization RNase A treatment which preserves hybrids but not single-stranded RNA molecules.

4. Hybridization

4.1 Generalities

A specific hybridization signal is obtained when the probe binds exclusively to the target sequence and when the resulting hybrids are stable. This requires the use of probes with high homology with the target, high stringency hybridization conditions, and a convenient temperature of hybridization. Influences of the composition of the buffer, probe concentration, and the time and temperature of hybridization on the intensity and specificity of the hybridization signal have been extensively detailed (6, 44–47) (see Chapter 1). We present here our routine conditions.

4.2 Hybridization solution

For our studies, the biotinylated double-stranded DNA probes and biotin labelled oligonucleotide probes are diluted at a final concentration ranging between 2–10 μg/ml in 50% formamide, 2 × SSC (1 × SSC = 0.15 M sodium chloride, 0.015 M sodium citrate), 10% dextran sulfate, and 400 μg/ml competitor tRNA or DNA. *Escherichia coli* DNA must be used as a competitor for ribosomal DNA probes because of the total absence of sequence homology with the human ribosomal genes. Salmon sperm DNA, which displays some sequence homology with portions of the ribosomal genes, is employed with the other DNA probes. tRNA is employed preferentially with oligonucleotide probes because it eliminates the necessity of a heat treatment of the hybridization solution prior to the hybridization step. 50 μl of hybridization solution is prepared from individual stock solutions and consists of:

(a) 25 μl formamide pH 6.8–7.2 from a deionized stock stored at 4°C in darkness.

(b) 10 μl dextran sulfate from a 200 μl stock solution at 50% in distilled water stored at 4°C.

(c) 5 μl SSC buffer from a 10 ml 20 × SSC stock solution stored at 4°C.

(d) 8 µl biotinylated probe from a stock, often at 60 µg/ml, occasionally less concentrated, stored at –20°C.

(e) 2 µl competitor nucleic acid from a 1 ml stock solution at 10 mg/ml stored at –20°C.

The hybridization solutions can be stored in the refrigerator in darkness for at least three years without variations in their efficiency.

4.3 Heat treatment of the hybridization solutions

The hybridization solutions containing double-stranded DNA probes and/or double-stranded DNA competitors must be submitted to a heat treatment just before use in order to denature the double-stranded molecules and allow subsequent base pairing of the resulting single-stranded probes with the complementary target sequences of the biological sample. Denaturation treatment of labelled double-stranded sequences is carried out as described in *Protocol 3*.

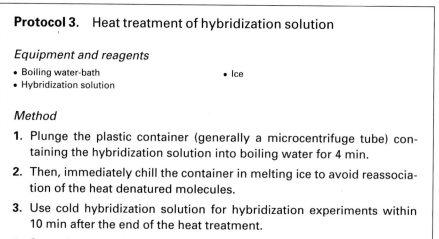

Protocol 3. Heat treatment of hybridization solution

Equipment and reagents
- Boiling water-bath
- Hybridization solution
- Ice

Method

1. Plunge the plastic container (generally a microcentrifuge tube) containing the hybridization solution into boiling water for 4 min.

2. Then, immediately chill the container in melting ice to avoid reassociation of the heat denatured molecules.

3. Use cold hybridization solution for hybridization experiments within 10 min after the end of the heat treatment.

4. Store the remaining solution at 4°C where it will remain useful for up to three years.

5. Heat treat the solution before each subsequent use.

4.4 Formation of hybrids on ultrathin sections

Hybridization is performed in order to anneal a labelled strand (the probe) with the unlabelled complementary strand of the biological sample. When performed on ultrathin sections of embedded material, the probe is accessible to only those complementary sequences which are exposed at the surface of the ultrathin section. Depending on the target under study, specific pretreatments of ultrathin sections (denaturation, enzymatic digestions) may be

Table 2. Successive steps for detecting nucleic acid targets at the surface of Lowicryl K4M sections using biotinylated double-stranded DNA probes[a]

Target	dsDNA + ssDNA	ssDNA	dsDNA	RNA
Pre-treatments of sections:				
Protease	Required	Required	Required	Recommended
RNase A	Required	Required	No	No
Nuclease S1	No	No	Required	No
DNase I	No	No	No	Required or omitted[b]
NaOH	Required	No	Required	No
Hybridization:				
Duration	90 min	90 min	90 min	Variable[c]
Temperature	37°C	37°C	37°C	Variable[c]
Post-hybridization (same for all targets):				
Washing	PBS, ~ 1 min			
Immunogold labelling	30 min			
Washing	PBS, dH₂O			
Staining	Uranyl acetate, 10 min			

[a] Abbreviatons used: ds: double-stranded; ss: single-stranded, dH₂O: distilled water.
[b] DNase I digestion is required for the specific detection of viral RNA in cells infected with herpes simplex virus type 1 and adenovirus because of the presence of viral single-stranded DNA portions.
[c] Hybridization conditions for detecting RNA target sequences are highly variable depending on the RNA sequence under study (see *Table 3*).

required prior to the hybridization step to obtain specific detection and a high hybridization signal (see *Table 2* and Section 5).

The hybridization step is similar whatever the pre-treatments of sections used although some conditions vary, including the temperature and the duration of the hybridization, and are adjusted empirically to maximize the hybridization signal. Hybridization is carried out at less than the T_m (melting temperature at which 50% of hybrids become dissociated), which depends upon several factors including the proportion of G + C in the probe, the length of the probe, the cation concentration, and the amount of formamide in the hybridization solution (see Chapter 1). We have chosen to keep constant the ionic strength and the formamide concentration and adjust stringency by varying the hybridization temperature. This is obtained by undertaking for each probe and each target (DNA, RNA) revealed with a given probe a series of experiments using different temperatures (37°C, 45°C, 55°C, 64°C) for the same period of time, generally 2 h. Then, a second series of experiments is undertaken by hybridizing at the selected temperature for different durations (60 min, 90 min, 3 h, 4 h).

As a guide, *Table 3* gives the conditions under which we obtained best results for detection with double-stranded DNA probes of: *Alu* sequences (31, 32), ribosomal genes (25, 36, 37), adenovirus and herpes simplex virus genomes (26, 28, 38), viral RNA (33, 48), U3 snoRNA (small nucleolar RNA)

Table 3. Hybridization conditions for the best detection of DNA or RNA target sequences at the surface of Lowicryl K4M sections using biotinylated probes

Probe	Target	Temperature	Duration
Double-stranded DNA probes:			
Ad5	Viral DNA	37 °C	60–90 min
HSV-1	Viral DNA	37 °C	60–90 min
Large portion of the rRNA gene	Ribosomal gene	37 °C	60–90 min
Alu elements	*Alu* DNA	37 °C	60–90 min
HFV	Parental RNA genome	37 °C	3.5 h
Ad5	Viral RNA	37 °C	3.5 h
HSV-1	Viral RNA	37 °C	3.5 h
5′ ETS	5′ ETS rRNA	55 °C	4 h
ITS1	ITS1 rRNA	55 °C	4 h
ITS2	ITS2 rRNA	55 °C	4 h
18S	18S rRNA	64 °C	4 h
Half end of rDNA gene	Mainly 28S rRNA	64 °C	4 h
U3	U3 snoRNA	37 °C	4 h
U1	U1 snRNA	64 °C	90 min
U2	U2 snRNA	64 °C	90 min
Oligonucleotide probe:			
poly(dT)	Cellular and viral poly(A)$^+$ RNA	37 °C	90 min

(29), ribosomal RNA (24, 36, 30), U1 and U2 snRNA (small nuclear RNA) (49); and the detection of poly(A) tails with an oligonucleotide probe (42).

4.5 Post-hybridization steps

The hybridization step is followed, first, by washing the samples under stringent conditions to remove non-specific background due to retention of the probe by physical entrapment and electrostatic attraction of charged groups; secondly, by immunogold labelling of hybrids; and thirdly, by staining of ultrathin sections prior to EM observation (see *Table 2*).

Post-hybridization washes in post-embedding EM *in situ* hybridization experiments are simpler and shorter than in optical experiments. Indeed, we perform washes at ambient temperature by rapid, successive passages of the hybridized grid on the surface of three drops of phosphate-buffered saline (PBS) for about 1 min total period. Without drying, the grid is processed for the detection of hybrids.

Hapten-containing hybrids are detected by either direct or indirect immunogold labelling. Since anti-biotin antibodies conjugated to gold particles are available commercially, we prefer to employ the one-step method which provides higher spatial resolution than the two-step method since the gold particles are separated from the hybrid targets only by the length of one immunoglobulin molecule. The direct immunogold labelling of hybrids con-

136

sists of incubation of the grid for 30 min at room temperature in specific anti-body conjugated to 10 nm gold particles. Ultrathin sections thereby labelled with gold particles are rinsed first by brief floating on three droplets of PBS and secondly, by rinsing in a jet of distilled water. The excess of water is blotted off the grids before air drying. Dry grids are stained for 10 min with 5% aqueous uranyl acetate solution, washed in a jet of distilled water, and air dried prior to transmission EM observation.

4.6 Hybridization protocol

The procedure which gives reproducible satisfactory data when *in situ* hybridization is performed on ultrathin sections of Lowicryl K4M-embedded material is as described in *Protocol 4* (see *Table 2*).

Protocol 4. Post-embedding *in situ* hybridization and detection of hybrids

Equipment and reagents

- Parafilm (American National Can)
- Grids bearing sections (see Section 2.4)
- Denatured hybridization solution (see Protocol 3)
- PBS
- Gold labelled anti-biotin antibody (Biocell Res. Lab)
- 5% aqueous uranyl acetate solution

Method

1. Denature the probe just before use by heating the hybridization solution (see Sections 4.2 and 4.3).

2. Distribute 1–2 μl drops of freshly denatured hybridization solution on a sheet of Parafilm placed in a wet chamber.

3. Float the grids, with or without pre-treatments (see Section 5), on the surface of the biotinylated probe-containing drops.

4. Place the wet chamber in an incubator for 1–4 h at a previously defined temperature (see Section 4.4).

5. Rinse the grids at room temperature for about 1 min by rapid passages over three drops of cold PBS distributed on a sheet of Parafilm (see Section 4.5).

6. Float the grid-mounted sections on 5 μl drops of anti-biotin antibody conjugated to gold particles, 10 nm in diameter, diluted 1:25 in PBS, at room temperature for 30 min (see Section 4.5).

7. Rinse the grids as above by floating them briefly on three droplets of cold PBS.

8. Wash the grids in a jet of distilled water.

9. Air dry the grids.

Protocol 4. *Continued*

10. Stain the grids for 10 min with 5% aqueous uranyl acetate solution before transmission EM observation.

11. Store the grids up to one year in darkness for further observations.

5. Specific preparations of target molecules prior to hybridization

5.1 Why and how does one improve the accessibility of the probe to the target?

Nucleic acid molecules are never naked in a cell, but always are associated with specific proteins which mask some portions of these molecules. The masked portions, therefore, cannot anneal to the probe, and must be unmasked by treatment of ultrathin sections with protease prior to the hybridization step. Such proteolytic treatment markedly improves the accessibility of the probe to its specific target (see *Table 2* and Section 5.1.1).

Elimination of proteins, however, is not always enough to allow the formation of hybrids because either the target is a double-stranded DNA molecule and, therefore not accessible at all, or the target, especially an RNA molecule, has a secondary structure with double-stranded portions. Denaturation of such double-stranded molecules or segments of molecules is mandatory before hybridization (see *Table 2*, Sections 5.1.2 and 5.1.3).

5.1.1 Elimination of the proteins of the ultrathin section

Removal of the proteins of the ultrathin sections is carried out by a protease digestion of ultrathin sections before hybridization as described in *Protocol 5*.

Protocol 5. Protease digestion of Lowicryl sections

Equipment and reagents
- Grids bearing sections
- Parafilm (American National Can)
- Incubator
- 0.2 mg/ml protease (or 1 mg/ml proteinase K) in water

Method

1. Prepare either a 0.2 mg/ml protease solution or a 1 mg/ml proteinase K solution in distilled water.

2. Distribute a 10 μl drop of proteolytic solution on a sheet of Parafilm in a wet chamber.

3. Float the grid on the surface of the enzyme-containing drop and put

the wet chamber in an incubator at 37 °C for either 15 min with protease or 1 h with proteinase K.

4. Rinse the grid rapidly at room temperature on three successive drops of distilled water distributed on a Parafilm sheet.

5. Wash the grid extensively in a jet of distilled water.

6. Air dry the grid.

7. The dry grid can be stored or immediately processed for hybridization with or without nuclease digestion(s) (Section 5.2) and denaturation treatment (Section 5.1.2).

Improvement of the accessibility of the target at the surface of the ultrathin section is not the only reason for elimination of the proteins from ultrathin sections. Non-specific retention of a probe to ultrathin sections may occur, at least in part, by electrostatic attraction between the probe and basic proteins in the section. This is suppressed by the protease pre-treatment of sections. The stickiness of proteins is highly variable. It has been reported that the histones of chromatin bind non-specifically to the probe only at late stages of the spermiogenesis of mammalian germ cells (50). In addition, we found in herpes simplex virus (HSV)-infected cells that the proteins of the viral core most avidly and non-specifically bind the HSV probe (27). A protease digestion of the ultrathin sections prior to hybridization is mandatory to completely suppress non-specific binding, as demonstrated by the total absence of labelling over the viruses in experiments localizing HSV RNA.

5.1.2 Denaturation of double-stranded DNA target molecules

Several different procedures are available to denature double-stranded DNA molecules in a section. We have tested some on Lowicryl K4M sections of herpes simplex virus type 1-infected cells fixed as described in Section 2.1 (27). We found that extensive denaturation of cellular and viral DNA only occurs following formaldehyde fixation, *and* after elimination of the proteins of the ultrathin section (see Section 5.1.1) whatever the DNA under study, viral or cellular.

Denaturation of the DNA in the ultrathin section with either 5 N HCl or 0.07 N NaOH prior to *in situ* hybridization with a specific viral probe results in a poor hybridization signal over both viruses and non-encapsidated viral genomes. Denaturation also is incomplete following heat treatment of sections prior to hybridization. Among the denaturation protocols tested, 0.5 N NaOH treatment was best for subsequent hybridization of both encapsidated and non-encapsidated herpes simplex genomes. Structural detail is well preserved during the 0.5 N NaOH treatment which also denatures genomes of adenovirus and cellular DNA, including ribosomal genes. The routine protocol is described in *Protocol 6*.

Protocol 6. Denaturation of DNA in Lowicryl sections

Equipment and reagents
- Grids bearing sections
- Parafilm (American National Can)
- 0.5 N NaOH

Method

1. Place the grid on a 10 μl drop of 0.5 N NaOH for 4 min at room temperature.
2. Rinse the grids by rapid passages on three 10 μl drops of distilled water.
3. Wash the grid in a jet of distilled water.
4. Air dry the grid for about 10 min.
5. Use the grid for *in situ* hybridization within 15 min after extensive washing in distilled water (see Section 4.6).

5.1.3 Denaturation of double-stranded RNA molecules

Some regions of RNA molecules form strong associations with other regions of the same molecules which produces three-dimensional folded structures. Such nucleotide pairing occurs particularly in snRNA molecules and in some viral transcripts including the adenovirus virus-associated (VA) RNA. The denaturation of the double-stranded portions of RNA sequences is not easy. It can be acheived by a heat and/or a formamide treatment of sections just before hybridization. Nevertheless, the hybridization signal is much more intense when target sequences are located in the normally single-stranded portions of the RNA molecule, as revealed by using several oligonucleotide probes for detecting U3 snoRNA in sections (51).

5.2 Why and how to perform nuclease digestion(s) prior to hybridization

When localizing DNA sequences, it is mandatory to digest ultrathin sections with pancreatic RNase A prior to hybridization in order to prevent con-comitant detection of related RNA sequences. When localizing RNA sequences, a prior DNase I digestion of ultrathin section is required only in those biological model systems which contain related single-stranded DNA sequences. As a guide (see *Table 2*), a DNase digestion must precede the detection of viral RNA in herpes simplex virus type 1-infected cells (HSV-1) (27) and adenovirus-infected cells (48) because of the presence of single-stranded portions in most HSV-1 genomes and the displacement of one viral DNA strand during the replication of adenoviral genomes. Conversely, the detection of cellular RNA does not require DNase pre-digestion of sections.

When localizing double-stranded DNA sequences only, a nuclease S1 pre-digestion of ultrathin sections is required to eliminate all single-stranded nucleic acid molecules, either RNA or DNA, in the ultrathin sections (52, 53). Following this treatment, double-stranded DNA molecules can anneal the probe provided that they have been denatured before the hybridization step (see Section 4.3).

Each nuclease treatment is performed at 37°C in a wet chamber for 1 h by floating the grids on 10 μl drops of freshly prepared enzyme-containing solution which are distributed on a sheet of Parafilm. This is followed by extensive washing of the grids in a jet of distilled water and air drying. Several nuclease treatments with or without protease pre-treatment (see Section 5.1.1) can be done successively on the same grid. Fortunately, dry digested grids can be stored for several days prior to subsequent denaturation and hybridization steps.

The nuclease solutions are prepared just before use as follows:

(a) RNase solution: dilute 1 mg/ml RNase A (from bovine pancreas; Worthington Biochemical Corp.) in 10 mM Tris–HCl buffer pH 7.3.

(b) DNase solution: dilute 1 mg/ml DNase I (Worthington Biochemical Corp.) in 10 mM Tris–HCl buffer pH 7.3, 5 mM $MgCl_2$, 2% RNasin ribonuclease inhibitor (Promega), 2 mM dithiothreitol. (RNasin suppresses possible contaminating RNase activity in the solution, and requires the presence of dithiothreitol to be active.)

(c) Nuclease S1 solution (final concentration of 1600 U/ml): dilute 1 μl 400 U/μl nuclease S1 (from Aspergillus oryzae) (Boehringer Mannheim) in 250 μl 1 × nuclease S1 buffer (10 × nuclease S1 buffer: 0.33 M sodium acetate pH 7.5, 0.5 M sodium chloride, 0.3 mM zinc sulfate). In our hands, lower concentrations of nuclease S1 do not hydrolyse all single-stranded nucleic acid molecules in Lowicryl sections of cells infected with adenovirus and fixed with formaldehyde (53).

6. Controls of the specificity of the detection

The specificity of the hybridization signal must be assessed by negative results following destruction of the target nucleic acid under study by enzymatic digestions (see Section 5.2). For example, an RNase A digestion of ultrathin sections prior to hybridization should suppress the detection of RNA sequences. Similarly, the detection of DNA molecules and single-stranded nucleic acid molecules should be abolished by pre-treatment of ultrathin sections with DNase I and nuclease S1, respectively. In addition, it must be emphasized that the use of RNase A in the washing steps before the hybrid detection step is mandatory to reduce the significant background which results from the use of riboprobes.

In addition to the nuclease digestions (see Section 5.2) which must prove that the hybridization depends on the presence of RNA or DNA, a series of other control experiments are required to assess the specificity of the hybridization signal, as follows:

(a) Protease digestion must be performed before the hybridization step to be sure that the probe binds only to nucleic acid and not to proteins of the ultrathin section (see Section 5.1.1).

(b) Hybridization either with irrelevant probes or by omitting the specific probe in the hybridization solution must produce negative results to ascertain that the detection system induced neither endogenous hapten (biotin) labelling nor non-specific background.

7. Routine protocols for specific detection of defined nucleic acid sequences and applications

Applications of post-embedding *in situ* hybridization are numerous. Protocols are now available for the specific detection of RNA and DNA elements with various configurations and origins, and have contributed to knowledge of the composition of some nuclear structures in normal and damaged cells:

(a) A new structure in normal cells, the so-called interchromatin granule-associated zone involved in the storage of U1 snRNA.

(b) The presence of polyadenylated RNA in clusters of interchromatin granules of normal and infected cells.

(c) The transient presence of viral RNA and ribosomal RNA in these nuclear structures following virus infection.

(d) The segregation of adenoviral DNA in well identifiable intranuclear structures depending on the level of involvement of viral genomes in replication and transcription.

The protocols that we have employed for detecting different types and forms of nucleic acid sequences contain some identical steps but vary in other crucial details depending on the nucleic acid sequence and biological system under study. We believe that the protocols described below may be applied in a rigorously controlled manner to a variety of investigations concerned with the structure–function relationships.

7.1 Specific detection of DNA sequences by using biotinylated double-stranded DNA probes

Most frequently, DNA target sequences are double-stranded. In certain models, however, portions of single-stranded DNA are normally present in the cell which can be specifically detected by means of appropriate *in situ* hybridization protocols. On the other hand, the normally double-stranded DNA segments also can be revealed provided that single-stranded nucleic acid sequences are removed prior to the hybridization step.

7.1.1 Concomitant detection of double- and single-stranded DNA sequences

When the DNA target molecules in the ultrathin section are either exclusively double-stranded such as the ribosomal genes (25, 36, 37) and *Alu* elements (31, 32) of the human genome, or they form a mixture of double-stranded and single-stranded segments such as viral DNA in HSV-1 and adenovirus-infected cells (26, 28, 38), a NaOH treatment of the ultrathin sections to denature the DNA duplexes must be performed prior to hybridization (*Figure 1*). The RNA molecules of the ultrathin section must be digested previously by a RNase A pre-treatment. In our model systems, additional elimination of the proteins of the ultrathin section markedly increases the hybridization signal over cellular and viral DNA. The successive steps for the best DNA detection are as described in *Protocol 7* and *Table 2*.

Protocol 7.   Successive steps for specific detection of total DNA (both double-stranded and single-stranded DNA)

Equipment and reagents

- Incubator
- Protease–RNase–NaOH pre-treated grids bearing sections
- Denatured hybridization solution
- Gold labelled anti-biotin antibody (Biocell Res. Lab)
- 5% aqueous uranyl acetate solution

Method

1. Eliminate the proteins (see Section 5.1.1 and *Protocol 5*) of the Lowicyl K4M section of formaldehyde-fixed material (see Sections 2.1 and 2.2).

2. Eliminate the RNA of the ultrathin section (see Section 5.3).

3. Denature the double-stranded DNA of the ultrathin section (see Section 5.1.2 and *Protocol 6*) and, in parallel, denature the double-stranded DNA of the hybridization solution (see Section 4.3 and *Protocol 3*).

4. Process as described in Section 4.6 (see *Protocol 4*) successively to form hybrids at the surface of the ultrathin section, to detect them by immunogold labelling, and to stain the ultrathin section prior to EM observation.

When applied to cells infected with HSV-1 and adenovirus type 5 (see *Table 3*), *in situ* hybridization revealed the total segregation of cellular DNA and progeny viral genomes in two concentric, juxtaposed compartments with-

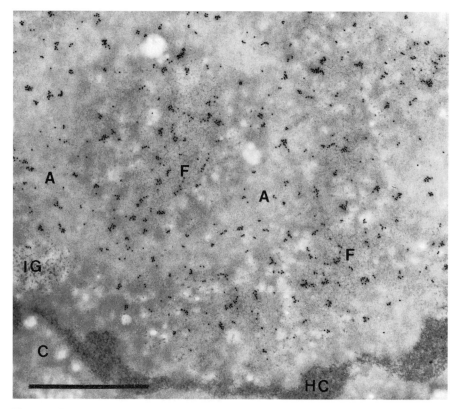

Figure 1. Detection of all adenoviral DNA by *in situ* hybridization. Adenovirus type 5-infected HeLa cell. Intermediate stage of nuclear modifications. Protease–RNase–NaOH pre-treatment of section prior to hybridization. Gold particles are present in the nucleus over the viral fibrillogranular network (F) and its enclosed ring-shaped viral ssDNA accumulation site (A). The host chromatin (HC), cellular cluster of interchromatin granules (IG), and cytoplasm (C) are devoid of labelling. Bar, 1 μm.

out interpenetration (*Figure 1*), even under experimental conditions which loosen the nucleoprotein fibres (31, 32). Such perfect compartmentalization of cellular and viral DNA is of great utility since it renders possible the clear identification of the DNA engaged in replication or transcription depending on its localization in the nucleus.

In situ hybridization also provides important data in less favourable model systems in which target molecules are much rarer. For example, it is possible to study the penetration of the parental viral genomes in the nucleus of the host cell and to follow their intranuclear behaviour before the initiation of viral genome replication (28). This has revealed that the HSV-1 parental genome penetrates the nucleus as a partially dissociated, swollen nucleoid which becomes completely dissociated in the interior of the nucleus. Then, intense replication of parental and subsequent progeny viral genomes induces the formation of a small, clear viral region which progressively becomes larger and pushes the cellular components to the nuclear border. The initiation of replication of adenovirus genomes also induces the progressive development of a centrally located viral region. However, this fibrillar region rapidly consists of different compartments resulting from a segregation of actively synthesized genomes, displaced DNA strands, and genomes in a resting state (54). The routine *in situ* hybridization protocol described above is applicable to many other viral infections.

With the same protocol but with a ribosomal DNA probe (*Table 3*), *in situ* hybridization has allowed the detection of ribosomal genes in nucleoli of normal cells (36) and in nucleoli modified by either an actinomycin D treatment (30) or HSV-1 infection (37). Interestingly, actinomycin D treatment induces the segregation of the usually intermingled granular component on the one hand, and dense fibrillar component with its enclosed fibrillar centres on the other hand, into three juxtaposed compartments, whereas HSV-1 infection results in a wide dispersion of the nucleolar components into granular spheres and contorted ribbons which partially surround rare and enlarged fibrillar centres. Despite marked variations in nucleolar architecture and in synthetic activity, ribosomal genes of the three systems which we tested, normal cells, actino-mycin D treated cells, and HSV-1-infected cells, always accumulate within the fibrillar centres and areas of the nucleolus-associated chromatin. The data suggest that fibrillar centres might contain the pool of ribosomal genes struc-turally ready for transcription whereas the nucleolus-associated condensed chromatin might contain the pool of unexpressed ribosomal genes.

7.1.2 Specific detection of single-stranded DNA sequences

The specific detection of single-stranded DNA sequences is achieved using the procedure described in Section 7.1 (see *Protocol 7*) except that the NaOH pre-treatment of ultrathin sections is omitted. As a result, only molecules which are normally single-stranded can bind the probe. Since RNA molecules are digested by the RNase A pre-treatment of ultrathin sections, only single-stranded DNA molecules can react with the probe (see *Table 2* and *Figure 2a*).

Single-stranded portions of DNA have been revealed by this procedure in progeny HSV-1 genomes before and after their insertion within the viral capsids (28). Therefore, the randomly distributed single-stranded portions observed in viral genomes extracted from mature viruses are not artefactually induced by biochemical techniques but are present *in vivo*. This peculiar feature of HSV-1 genomes, therefore, necessitates a DNase pre-treatment of the sections prior to detection by *in situ* hybridization of the viral RNA in infected cells (see Section 7.2.2).

Single-stranded DNA is known to appear in nuclei of cells lytically infected by adenovirus due to the mode of replication of the adenoviral genomes (54). One strand is replicated in a continuous fashion whereas the other is displaced and replicated later, also in a continuous fashion. The intense replicative

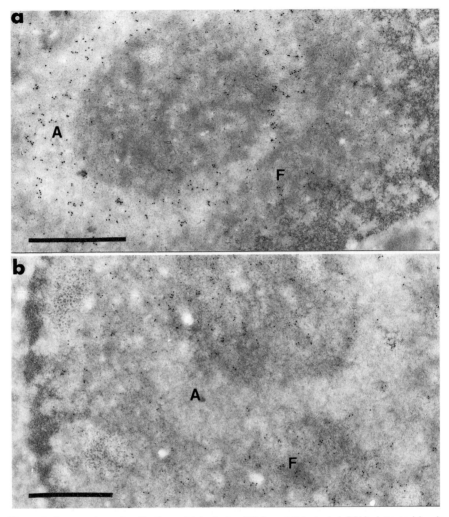

Figure 2. Intranuclear distribution of (a) viral ssDNA and (b) viral dsDNA detected by *in situ* hybridization. Adenovirus type 5-infected HeLa cell. Intermediate stage of nuclear modifications. (a) Protease–RNase pre-treatment of section prior to hybridization. In the absence of NaOH denaturation, only viral ssDNA is revealed. Gold particles accumulate over the ring-shaped viral ssDNA accumulation site (A) whereas a few particles are scattered in the surrounding fibrillogranular network (F). (b) Protease–nuclease S1–NaOH pre-treatment of section prior to hybridization. Only viral dsDNA is revealed. Gold particles are present over the viral fibrillogranular network (F) which fills the nucleus and are absent from the enclosed viral ssDNA accumulation site (A). Bars, 1 μm.

activity of adenoviral genomes induces the formation of many viral single-stranded DNA strands, the distribution of which in the nucleus is unknown. Non-isotopic *in situ* hybridization at the EM level was the first procedure to reveal that the viral single-stranded DNA molecules aggregate to form a virus-induced, morphologically easily identifiable, compact fibrillar structure (38, 54) (*Figure 2a*). It must be kept in mind, therefore, that in studies bearing on the intranuclear localization of adenoviral RNA molecules, total elimination of the viral single-stranded DNA is mandatory prior to the hybridization step (see Section 7.2.2).

7.1.3 Specific detection of double-stranded DNA sequences

In model systems containing both single DNA strands and duplex DNA molecules, it is informative to determine the relative amounts of single-stranded and double-stranded DNA molecules in the different substructures of the nucleus. In Section 7.1.2, conditions for the exclusive detection of normally single-stranded DNA molecules are described. Here, we describe experimental conditions for the exclusive detection of double-stranded DNA molecules (see *Table 2* and *Figure 2b*). This procedure uses nuclease S1 to eliminate all single-stranded nucleic acid molecules, both RNA and DNA, in the ultrathin sections. A protease pre-treatment of the ultrathin sections is required because specific DNA binding proteins protect single-stranded DNA. We process the sections as described in *Protocol 8*.

Protocol 8.   Protocol for specific detection of double-stranded DNA in Lowicryl sections

Equipment and reagents

- Incubator for hybridization
- Grids bearing sections
- Protease (or proteinase K)
- Nuclease S1 from Aspergillus oryzae (commercial stock solution at 400 U/ml) (Boehringer Mannheim)
- 10 × nuclease S1 buffer: 0.33 M sodium acetate, 0.5 M sodium chloride, 0.3 mM zinc sulfate pH 7.5
- 0.5 M NaOH
- Gold labelled anti-biotin antibody (Biocell Res. Lab)

Method

1. Digest proteins by a protease treatment (see Section 5.1.1 and *Protocol 5*) of Lowicryl K4M sections of formaldehyde-fixed material (see Sections 2.1 and 2.2).

2. Digest single-stranded nucleic acid molecules of the ultrathin section with nuclease S1 (1600 U/ml for 1 h at 37^°C) (see Section 5.2).

3. Denature double-stranded DNA of the ultrathin section by a NaOH treatment (0.5 M NaOH for 4 min at room temperature) (see Section 5.1.2 and *Protocol 6*) and, in parallel denature the double-stranded DNA of the hybridization solution by a 4 min heat treatment (see Section 4.3 and *Protocol 3*).

4. According to *Protocol 4* (Section 4.6), form hybrids at the surface of the ultrathin section, detect the hybrids by immunogold labelling, and then stain the ultrathin section prior to EM observation.

It is imperative to ascertain the total removal of all single-stranded nucleic acid sequences by undertaking hybridization of the probe in the absence of a previous denaturation treatment of ultrathin sections, i.e. by omitting the first part of step 3. There should be no labelling due to the enzymatic elimination of single-stranded RNA and DNA, and the inaccessibility of the probe to the non-denatured double-stranded DNA. If there is residual labelling, a higher concentration of nuclease S1 and/or more prolonged incubation in the presence of the enzyme should be employed.

The application of this procedure to adenovirus-infected cells together with a specific viral probe has allowed the clear demonstration of the total absence of viral double-stranded DNA molecules in the viral single-stranded DNA accumulation sites (compare *Figure 2a* and *Figure 2b*), and therefore definitely demonstrated that these structures are not sites of replication for viral genomes (52, 53, 56). This was in question because of the concomitant presence in the viral single-stranded accumulation sites of large amounts of one of the viral proteins essential for viral genome replication.

7.2 Specific detection of RNA sequences by using biotinylated probes

We have used non-isotopic EM *in situ* hybridization with a view to:

(a) Localizing total viral RNA and defined families of viral RNA in infected cells and in cells transfected with a viral gene.

(b) Studying the maturation of pre-ribosomal transcripts by localizing primary transcripts and their intermediates of maturation in normal and virus-infected cells.

(c) Studying the maturation of cellular and viral transcripts by localizing the small nuclear RNA (snRNA) and poly(A) tails.

(d) Localizing parental viral genomes at the early phase of single-stranded RNA virus infection (see *Table 3*).

7.2.1 Generalities and protocol

Except for the probe used for the detection of the poly(A) tails which was a synthetic poly(dT) probe biotinylated by 3' end tailing procedure (see Section 3.2.2), all other probes that we have used were double-stranded DNA biotinylated by nick translation, and therefore had to be denatured before use (see Section 4.3). Since the target RNA sequences are single-stranded, alkali treatment of ultrathin sections cannot be used. As a result, a high signal is obtained over Lowicryl K4M sections of material fixed with either formaldehyde or glutaraldehyde (see Section 2.1). Both fixations allow efficient enzymatic digestions of sections prior to the hybridization step. These digestions render the signal specific and frequently increase the accessibility of the probe to the target sequences (see Sections 5.1.1, 5.2, and 6) (compare *Figure 3a* and *3b*).

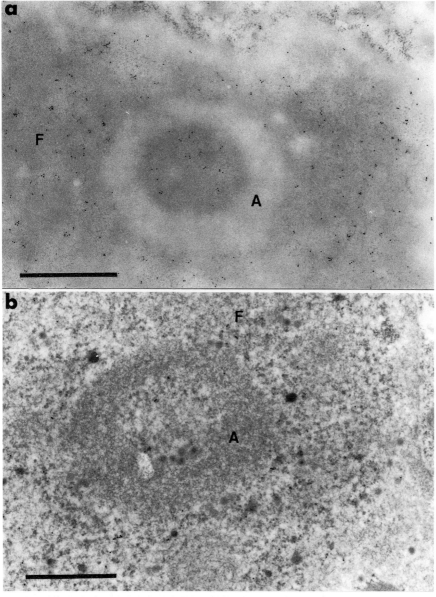

Figure 3. Intranuclear distribution of viral RNA detected by *in situ* hybridization performed in the (a) absence or (b) presence of cellular proteins in the section. Adenovirus type 5-infected HeLa cell. Intermediate stage of nuclear modifications. (a) Protease pre-treatment of the section prior to hybridization. Gold particles which localize viral RNA are numerous over the viral fibrillogranular network (F) and are absent over the enclosed ring-shaped viral ssDNA accumulation site (A). (b) Without protease pre-treatment of the section, the labelling of the fibrillogranular network (F) is slight. Once again, the ring-shaped viral ssDNA accumulation site (A) is devoid of gold particles. Bars, 1 µm.

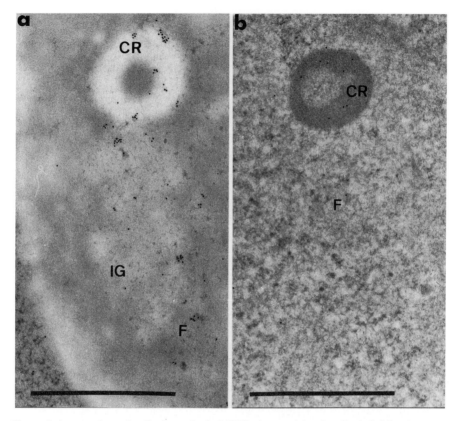

Figure 4. Intranuclear distribution of viral RNA detected by *in situ* hybridization performed in the (a) absence or (b) presence of cellular proteins in the section. Adenovirus type 5-infected HeLa cell. Intermediate stage of nuclear modifications. (a) Protease pretreatment of section prior to hybridization. Gold particles which localize viral RNA are numerous over the viral compact ring (CR) and fibrillogranular network (F), and over the cellular cluster of interchromatin granules (IG). (b) Without protease pre-treatment of section, the labelling of the viral compact ring (CR) persists whereas that of the viral fibrillogranular network (F) is minimal. Bars, 1 µm.

It must be pointed out that the degree of accessibility of a defined RNA target sequence may vary in different regions of the same cell. For example, a high level of detection is obtained for adenoviral RNA which is inserted within virus-induced structures which assume the shape of a compact ring even in the presence of the proteins of the ultrathin section (48) (compare *Figure 4a* and *4b*). In contrast, a good hybridization signal over the nuclear compartment containing transcribing adenoviral genomes requires a protease pre-treatment (compare *Figure 3a* and *3b*). Therefore, it is recommended that the accessibility of the probe to each cell substructure be ascertained by comparing the intensity of the hybridization signal in the presence and in the absence of proteins. In addition, it must be kept in mind that some proteins, for example the nucleoid proteins of the HSV-1 viruses (27), may stick to the probe and give false positive results. This does not occur following a protease pre-treatment of the ultrathin sections.

Another crucial point is the presence of single-stranded DNA segments in some model systems, such as cells infected with HSV-1 and adenovirus (see Section 7.1.2), which are revealed concomitantly with the related RNA sequences. The specific detection of the viral RNA sequences in such biological materials imperatively requires the previous elimination of DNA by a DNase I treatment.

Specific detection of RNA sequences is achieved as described in *Protocol 9* (see *Table 2*).

Protocol 9. Successive steps for specific detection of RNA molecules in Lowicryl sections

Equipment and reagents

- Incubator for hybridization
- Grids bearing sections with or without protease and/or DNase I digestion
- Denatured hybridization solution
- Gold labelled anti-biotin antibody (Biocell Res. Lab)

Method

1. Use ultrathin sections of formaldehyde- or glutaraldehyde-fixed material (see Section 2.1).

2. Depending on the model system under study, if required perform protease digestion (Section 5.1.1 and *Protocol 5*) and/or DNase I digestion of ultrathin sections (Section 5.2).

3. Heat treat the hybridization solution if it contains double-stranded DNA molecules (see Section 4.3 and *Protocol 3*).

4. Incubate the grids in the presence of the probe at a selected temperature and duration depending on the GC content of the probe (see Section 4.4).

5. Reveal the hybrids and stain the grids before observation as described in Section 4.6 (see *Protocol 4*, steps 5–11).

7.2.2 Detection of viral RNA in infected cells

Viral RNA molecules in DNA virus-infected cells consist of primary transcripts, their various intermediates of maturation, and the cytoplasmic messenger RNA molecules. When infectious viruses are RNA viruses, the cells also contain genomic RNA molecules.

By *in situ* hybridization with biotinylated double-stranded DNA probes, we have studied the expression of viral genes and the maturation of transcripts in

HSV-1 and adenovirus type 5-infected cells, i.e. in DNA virus-infected cells (33, 34, 39). We have also localized viral RNA genomes in cells infected with human foamy virus (HFV), a spumaretrovirus, at the early stage of the infectious cycle that just precedes the intranuclear translocation of the pre-integration complex (57) (see Section 4.4, *Table 3* for conditions). DNase digestion of ultrathin sections was required and protease digestion is recommended for detection of viral RNA in cells infected with HSV-1 and adenovirus (see Section 7.2.1) (*Figure 3a*). Enzymatic digestions were without effect on the level of detection of parental viral RNA genomes in early HFV-infected cells (*Figure 5*).

Regarding the localization of viral RNA in DNA virus-infected cells, hybridization signals are detected as expected over the ribosome-containing cytoplasmic areas and the intranuclear compartment in which transcribing viral genomes accumulate (*Figure 3a*). Surprisingly, viral RNAs synthesized from HSV-1 and adenovirus genomes also are present in the cellular clusters of interchromatin granules (*Figure 4a*) together with poly(A) tails. This exciting *in situ* hybridization result suggests that clusters of interchromatin granules might be transient storage sites and/or a sorting site for messenger RNA molecules before their migration to the cytoplasm, although we cannot exclude the possibility that clusters of interchromatin granules might be a degradation site for useless or nonsense viral messenger RNA (33, 34, 39, 40). Adenovirus infection involves a further specific localization of viral RNA. A

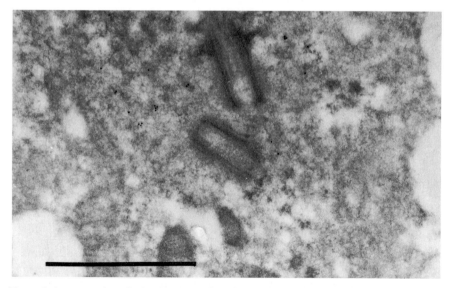

Figure 5. Intracytoplasmic localization of incoming viral genomes detected by *in situ* hybridization. Human foamy virus (HFV)-infected MRC5 cell. Early stage of infection. No pre-treatment of section. Gold particles which localize RNA HFV genomes are scattered within the pericentriolar material. Bar, 1 μm.

genomic probe reveals that adenoviral RNA molecules are present in compact rings (*Figure 4a* and *4b*), intranuclear virus-induced structures which are contiguous to the clusters of interchromatin granules and to the compartment of transcribing viral genomes. Since compact rings also contain spliceosome components but are devoid of poly(A) tails, we suggest that the unused portions of the primary adenoviral late transcripts might accumulate in the compact rings subsequently to be destroyed and the spliceosome components released for further use.

Regarding the early stage of HFV infection, we observe that the parental viral RNA genomes, together with viral proteins, accumulate around the centrosome of the infected cell, probably to constitute the pre-integration complex which will be subsequently translocated into the nucleus (57) (*Figure 5*). *In situ* hybridization was the first process to reveal an interaction between a retroviral pre-integration complex and the centrosome of the host cell.

Post-embedding EM *in situ* hybridization is also helpful in studies of cells transfected with a defined gene. We chose to localize the RNA product of the Us11 gene of the HSV-1 genome following transient transfection of HeLa cells because the resulting 21 kDa viral protein accumulates within the nucleoli of lytically-infected cells and transfected cells. We observe that large amounts of Us11 RNA and poly(A)$^+$ RNA accumulate within the clusters of interchromatin granules following transfection but never are found in the 21 kDa-containing nucleoli (40). This approach therefore is fruitful for studying the interaction between defined families of messenger RNA molecules and specific nuclear domains or structures, and in particular, for improving our knowledge of the functional significance of their retention within the cellular clusters of interchromatin granules.

7.2.3 Detection of small nuclear RNA and poly(A)$^+$ RNA

Cellular and viral pre-messenger RNA are modified by a series of post-transcriptional processing events which include capping, polyadenylation, and splicing. Splicing of pre-messenger RNA requires spliceosomes, cellular ribonucleoprotein particles which result from the assembly of pre-spliceosome particles containing distinct small nuclear RNA (snRNA) molecules (U1, U2, U4/U6, or U5 snRNA). *In situ* hybridization has enabled the localization of the different kinds of snRNA and the poly(A)$^+$ RNA molecules by using biotinylated probes (see Section 4.4, *Table 3* for conditions).

In the absence of virus infection, some distinct intranuclear structures accumulate snRNA. For example, clusters of interchromatin granules contain large amounts of U1 snRNA and U2 snRNA whereas coiled bodies contain large amounts of U2 snRNA but only traces of U1 snRNA. Interestingly, EM *in situ* hybridization has revealed the accumulation of U1 snRNA and the total absence of U2 snRNA in a newly described structure that we have designated an 'interchromatin granule-associated zone' (49). These results suggest that different intranuclear structures including coiled bodies and interchromatin granule-associated zones might be involved in the formation and/or storage of pre-spliceosome particles whereas the clusters of inter-chromatin granules might be the site of their final assembly.

A few hours after exposure of cells to HSV-1 and adenovirus type 5, coiled bodies and interchromatin granule-associated zones disappear. Only the clusters of interchromatin granules and their unmodified snRNA content persist. However, clusters of interchromatin granules contain significant amounts of viral RNA molecules and poly(A) tails (see Section 7.2.2) (33, 39, 48). Since poly(A)$^+$ RNA molecules are present, although at a reduced level, in the clusters of interchromatin granules of non-infected cells (42), the involvement of the clusters of interchromatin granules in the storage, sorting, and/or degra-dation of viral messenger RNA suggested by the *in situ* hybridization data might be general and relate to the fate of cellular mRNA molecules.

7.2.4 Detection of ribosomal RNA transcripts and their intermediates of maturation

By using probes mapping for selected portions of the human ribosomal genes, it is possible to detect either ribosomal RNA or its different intermediates of maturation under either normal conditions of growth or following alterations of nucleolar metabolism by actinomycin D treatment and DNA virus in-fection (see *Table 3*). For example, following the use of a 5′ ETS (5′ external transcribed spacer) rDNA probe (36), it is possible to detect only the growing and primary transcripts, since the 5′ ETS rRNA portion is the first to be removed from the just-finished primary transcript leading to the formation of a 41S pre-rRNA. Since the early processing reaction involves the presence of a ribonucleoprotein containing the small nucleolar RNA (snoRNA) U3, the use of a U3 DNA probe also allows the detection of precursors and nascent rRNA molecules (29). As a result, U3 snoRNA and 5′ ETS RNA are found over the dense fibrillar component of the nucleolus including its interface with the enclosed fibrillar centres. 5′ ETS rRNA also is present at the border of the fibrillar centres, a result which strongly suggests that only the ribosomal genes located at this site are actively engaged in transcription. On the other hand, U3 snoRNA, which is expected to play a role not only in the initial cleavage of rRNA primary transcript but also in the subsequent rRNA processing steps, is also found in the granular component of the nucleolus co-localizing with partially processed pre-rRNA molecules.

Actinomycin D treatment of cells, which is known to block the formation of new transcripts but is without effect on the processing of the previously synthesized pre-rRNA molecules, results in the persistence of U3 snoRNA in the rRNA-depleted dense fibrillar component of the segregated nucleoli (30). These data indicate that those U3 snoRNA molecules which are not engaged in rRNA processing are stored in the dense fibrillar component of the nucleolus. By using in parallel probes for other portions of the ribosomal gene, such as the ITS (internal transcribed spacer) 1, ITS 2, 18S, and 28S portions, EM *in situ* hybridization has proven to be a powerful approach for analysing the molecular organization of the nucleolus.

Infection of cells with HSV-1 results in marked morphological changes in the nucleolus. However, rDNA probes label the virus-altered nuclei in the same way as for non-infected cells with a few differences:

(a) rRNA molecules are numerous in the clusters of interchromatin granules and the intranuclear viral region.

(b) Larger amounts of 18S rRNA are retained in the granular component of the virus-modified nucleoli (58).

The virus-induced retention of rRNA molecules in some nuclear structures of infected cells, including the clusters of interchromatin granules, suggests that the latter might be implicated in the regulation of the export of the ribosomal subunits towards the cytoplasm. DNA virus infections are promising model systems for studying the regulation of the nucleo–cytoplasmic transport of RNA molecules, cellular and viral messenger RNA (Sections 7.2.2 and 7.2.3), as well as ribosomal subunits (this section).

8. Double detections

The possibility of using different sizes of gold particles allows the simultaneous detection on the same grid of either two distinct nucleic acid sequences or a nucleic acid sequence and an antigen. Detection is achieved using either direct or indirect methods alone, or a combination of direct and indirect methods.

8.1 Simultaneous detection of two distinct nucleic acid sequences

Simultaneous detection of two distinct nucleic acid sequences is possible provided that:

(a) The two probes are labelled with different haptens, for example, one probe with biotin and the other with digoxigenin.

(b) The two target sequences are either RNA or DNA. The two target sequences cannot be RNA for one and DNA for the other since the detection of DNA sequences requires RNase and NaOH pre-treatments, both of which hydrolyse RNA (see Sections 5.1.2 and 5.2), whereas detection of RNA sequences in some model systems requires DNase pre-treatment in order to eliminate single-stranded DNA sequences (see Section 7.2.1).

When the two distinct nucleic acid sequences require identical conditions of temperature for the hybridization step (see Sections 4.4 and 4.6) the grids are incubated in a mixture of probes. If hybridization temperatures vary, grids are incubated, first, in the presence of the probe which requires the higher temperature of hybridization, then, after a rapid wash over drops of PBS, the grid is reincubated over the other hybridization solution at the selected lower temperature. The detection of the two kinds of hybrids is performed as described below for the the simultaneous detection of a nucleic acid sequence and an antigen (see Section 8.2 from step (b) to the end). We plan to use this still rarely used procedure at the ultrastructural level (6, 59) to detect simultaneously in the same cell section the products of adenoviral genes transcribed by RNA polymerases II and III.

8.2 Simultaneous detection of a nucleic acid sequence and an antigen

Provided that detection of a nucleic acid sequence does not require elimination of the proteins of the ultrathin section, it is possible to detect on the same grid a defined nucleic acid sequence by *in situ* hybridization and a protein by immunogold labelling (39, 60) as follows:

(a) Incubate the grid in the presence of the hybridization solution for a defined duration and at a defined temperature (see Section 4.4).

(b) Rapidly wash the grid (see Section 4.5).

(c) Without drying, incubate the grid in the presence of the antibody at ambient temperature for a selected duration.

(d) Subsequently, either:

  (i) Without drying, incubate the grid for 30 min at ambient temperature in a mixture of gold labelled anti-hapten antibody and gold labelled anti-species antibody in order to specifically label with gold particles of different sizes the hybrids and the antigen–antibody complexes formed at the surface of the ultrathin sections.

  (ii) When hybrids must be revealed by the indirect immunogold procedure, first incubate the grid in the presence of the unlabelled anti-hapten antibody, for example rabbit anti-biotin antibody, for 30 min at ambient temperature, then over a mixture of rabbit anti-IgG conjugated to 5 nm gold particles and anti-species antibody (other than rabbit) conjugated to 10 nm gold particles.

(e) Finally, stain the grid prior to observation with the transmission electron microscope.

9. Conclusions

We have developed post-embedding *in situ* hybridization protocols allowing the rapid and reproducible localization of defined nucleic acid sequences of different configurations and origins, cellular or viral, DNA or RNA, single- or double-stranded molecules. The major interest rests in the clear identification of the cell structure or domain containing the target sequence under study. These *in situ* hybridization protocols used in parallel or simultaneously with other refined EM techniques have revolutionized the study of the nucleus and allowed us to relate ultrastructure to the synthetic activities of genomes and the molecular aspects of gene expression under both normal conditions of cell growth and following alterations of the cellular metabolism by drug, viral infection, and expression of a foreign gene.

Acknowledgements

The author is very indebted to Professor E. H. Leduc (Brown University, Providence, RI, USA) for critical reading of the manuscript. This work is supported by general grants from the Centre National de la Recherche Scientifique and special grants from the Association pour la Recherche sur le Cancer (ARC, Villejuif, France) and the Ligue Nationale contre le Cancer. F.P-D is a member of the Institut National de la Santé et de la Recherche Médicale.

References

1. John, H.A., Birnstiel, M.L., and Jones, K.W. (1969). *Nature*, **223**, 582.
2. Pardue, M.L. and Gall, J.G. (1969). *Proc. Natl. Acad. Sci. USA*, **64**, 600.
3. Binder, M., Tourmente, S., Roth, J., Renaud, M., and Gehring, W.J. (1986). *J. Cell Biol.*, **102**, 1646.
4. Narayanswami, S. and Hamkalo, B.A. (1987). In *Electron microscopy in molecular biology: a practical approach* (ed. J. Sommerville and U. Scheer), p. 215. IRL Press, Oxford.
5. O'Reilly, M.M., French, S.L., Sikes, M.L., and Miller, O.L., Jr. (1994). *Chromosoma*, **103**, 122.
6. Morel, G. (1993). In *Hybridization techniques for electron microscopy* (ed. G. Morel), p. 163. CRC Press, Boca Raton.
7. Trembleau, A. (1993). In *Hybridization techniques for electron microscopy* (ed. G. Morel), p. 139. CRC Press, Boca Raton.
8. Fournier, J.G. and Escaig-Haye, F. (1993). In *Hybridization techniques for electron microscopy* (ed. G. Morel), p. 243. CRC Press, Boca Raton.
9. Puvion-Dutilleul, F. (1993). In *Hybridization techniques for electron microscopy* (ed. G. Morel), p. 269. CRC Press, Boca Raton.
10. Puvion-Dutilleul, F., Pichard, E., Laithier, M., and Leduc E.H. (1987). *J. Histochem. Cytochem.*, **35**, 635.
11. Roth, J., Bendayan, M., Carlemalm, E., and Villiger, W. (1981). *J. Histochem. Cytochem.*, **29**, 663.
12. Scala, C., Cenacchi, G., Ferrari, C., Pasquinelli, G., Preda, P., and Manara, G.C. (1992). *J. Histochem. Cyrochem.*, **40**, 1799.
13. Bogers, J.J., Nibbeling, H.A., Deelder, A.M., and Vanmarck, E.A. (1996). *J. Histochem. Cytochem.*, **44**, 43.

14. Cenacchi, G., Musiani, M., Gentilomi, G., Righi, S., Zerbini, M., Chandler, J.G., *et al.* (1993). *J. Submicrosc. Cytol. Pathol.*, **25**, 341.
15. Bendayan, M. (1984). *J. Electron Microsc. Tech.*, **1**, 243.
16. Escaig-Haye, F., Grigoriev, V., Peranzi, G., Lestienne, P., and Fournier, J.G. (1991). *J. Cell Sci.*, **100**, 851.
17. Langer-Safer, P.R., Levine, M., and Ward, D.C. (1982). *Proc. Natl. Acad. Sci. USA*, **79**, 4381.
18. Dadoune, J.P., Siffroi, J.P., and Alfonsi, M.F. (1994). *Cell Tissue Res.*, **278**, 611.
19. De Furtos, R., Kimura, K., and Peterson, K.R. (1989). *Trends Genet.*, **5**, 366.
20. Yi, J., Michel, O., Sassy-Prigent, C., and Chevalier, J. (1995). *J. Histochem. Cytochem.*, **43**, 801.
21. Niedobitek, G., Finn, T., Herbst, H., and Stein, H. (1989). *Am. J. Pathol.* **134**, 633.
22. Yao, C.H., Kitazawa, S., Fujimori, T., and Maeda, S. (1993). *Biotech. Histochem.*, **68**, 169.
23. Multhaupt, H.A.B., Rafferty, P.A., and Warhol, M.J. (1992). *Lab. Invest.*, **67**, 512.
24. Pierron, G. and Puvion-Dutilleul, F. (1993). *Exp. Cell Res.*, **208**, 509.
25. Puvion-Dutilleul, F. and Pierron, G. (1992). *Exp. Cell Res.*, **203**, 354.
26. Puvion-Dutilleul, F. and Puvion, E. (1989). *Eur. J. Cell Biol.*, **49**, 99.
27. Puvion-Dutilleul, F. and Puvion, E. (1991). *J. Electron Microsc. Tech.*, **18**, 336.
28. Puvion-Dutilleul, F., Pichard, E., Laithier, M., and Puvion, E. (1989). *Eur. J. Cell Biol.*, **50**, 187.
29. Puvion-Dutilleul, F., Mazan, S., Nicoloso, M., Christensen, M.E., and Bachellerie, J.P. (1991). *Eur. J. Cell Biol.*, **56**, 178.
30. Puvion-Dutilleul, F., Mazan, S., Nicoloso, N., Pichard, E., Bachellerie, J.P., and Puvion, E. (1992) .*Eur. J. Cell Biol.*, **58**, 149.
31. Besse, S. and Puvion-Dutilleul, F. (1994). *Chromosome Res.*, **2**, 123.
32. Puvion-Dutilleul, F. and Besse, S. (1994). *Chromosoma*, **103**, 104.
33. Besse, S., Vigneron, M., Pichard, E., and Puvion-Dutilleul, F. (1995). *Gene Expression*, **4**, 143.
34. Puvion-Dutilleul, F., Bachellerie, J.P., Visa, N., and Puvion, E. (1994). *J. Cell Sci.*, **107**, 1457.
35. Dakashinamurti, K. and Mistry, S.P. (1963). *J. Biol. Chem.*, **238**, 294.
36. Puvion-Dutilleul, F., Bachellerie, J.P., and Puvion, E. (1991). *Chromosoma*, **100**, 395.
37. Besse, S. and Puvion-Dutilleul, F. (1996). *Eur. J. Cell Biol.*, **71**, 33.
38. Puvion-Dutilleul, E. and Puvion, E. (1990). *Eur. J. Cell Biol.*, **52**, 379.
39. Besse, S. and Puvion-Dutilleul, F. (1995). *Gene Expression*, **5**, 79.
40. Besse, S., Diaz, J.J., Pichard, E., Kindbeiter, K., Madjar, J.J., and Puvion-Dutilleul, F. (1996). *Chromosoma*, **104**, 434.
41. Tani, T., Berby, R.J., Hiraoka, Y., and Spector, D.L. (1995). *Mol. Biol. Cell*, **6**, 1515.
42. Visa, N., Puvion-Dutilleul, F., Harper, F., Bachellerie, J.P., and Puvion, E. (1993). *Exp. Cell Res.*, **208**, 19.
43. Raska, I., Dundr, M., Koberna, K., Melcak, I., Risueno, M.C., and Török, I. (1995). *J. Struct. Biol.*, **114**, 1.
44. McFadden, G.I. (1995). *Methods in Cell Biology*, **49**, 165.
45. Binder, M. (1992). In *In situ hybridization: a practical approach* (ed. D.G. Wilkinson), p. 105. IRL Press, Oxford.

46. Polak, J.M. and McGee, J.O.D. (1990). *In situ hybridization. Principles and practices* (ed. J.M. Polak and J.O.D. McGee), pp. 1–247. Oxford University Press, Oxford.
47. Wilkinson, D.G. (1992). In *In situ hybridization: a practical approach* (ed. D.G. Wilkinson), p. 1. IRL Press, Oxford.
48. Puvion-Dutilleul, F., Roussev, R., and Puvion, E. (1992). *J. Struct. Biol.*, **108**, 209.
49. Visa, N., Puvion-Dutilleul, F., Bachellerie, J.P., and Puvion; E. (1993). *Eur. J. Cell Biol.*, **60**, 308.
50. Galdieri, M. and Monesi, V. (1973). *Boll. Zool.*, **40**, 411.
51. Puvion-Dutilleul, F., Puvion, E., and Bachellerie, J.P. (1997). *Chromosoma*, **105**, 496.
52. Puvion-Dutilleul, F. and Pichard, E. (1992). *Biol. Cell*, **76**, 139.
53. Puvion-Dutilleul, F. (1993). *Microsc. Res. Tech.*, **25**, 2.
54. Puvion-Dutilleul, F. and Puvion, E. (1990). *J. Struct. Biol.*, **103**, 280.
55. Lechner, R.L. and Kelly, T.J., Jr. (1977). *Cell*, **12**, 1007.
56. Besse, S. and Puvion-Dutilleul, F. (1994). *Eur. J. Cell Biol.*, **63**, 269.
57. Saïb, A., Puvion-Dutilleul, F., Schmid, M., Périès, J., and de Thé, H. (1997). *J. Virol.*, **71**, 1155.
58. Besse, S. and Puvion-Dutilleul, F. (1996). *J. Cell Sci.*, **109**, 119.
59. McFadden, G., Bönig, I., and Clarke, A. (1990). *Trans. R. Microsc. Soc.*, **1**, 683.
60. Mertani, H.C., Watters, M.J., Jambou, R., Gossard, F., and Morel, G. (1994). *Neuroendocrinology*, **59**, 483.

Detection of genomic sequences by fluorescence *in situ* hybridization to chromosomes

LYNDAL KEARNEY

1. Introduction

Since the first report over ten years ago (1), fluorescence *in situ* hybridization (FISH) has fulfilled its' promise as a powerful tool in genome analysis. The type of genomic sequence that can be used as a probe now ranges from small sequence tagged sites (STSs) to whole genomes (reviewed in ref. 2). With the advent of new labelling and imaging systems (3, 4), the sensitivity now approaches that of earlier isotopic methods, with the ability to detect hybridized sequences < 1 kb in size. Recently, the possibility of 24 colour FISH has been achieved, enabling the simultaneous analysis of the entire chromosome complement in a single multicolour hybridization experiment (5, 6). The chromosomal targets for FISH are equally varied, from metaphase chromosomes to extended strands of DNA (reviewed in ref. 7). High-resolution metaphase chromosomes can be used to resolve two sequences approximately one megabase (Mb) apart, while interphase mapping allows the resolution of sequences 100 kb apart (8). Techniques using extended DNA fibres have revolutionized fine mapping, allowing the ordering and orientation of sequences 1 kb apart (9–12).

FISH technology can be applied to an enormous variety of clinical and research situations. Applications include:

- gene mapping
- the identification of numerical and structural chromosome abnormalities
- interphase cytogenetics
- the identification of the human content of somatic cell hybrids
- the identification of new regions of amplification or deletion
- positional cloning
- comparative cytogenetics

This chapter will concentrate on FISH as an aid to gene mapping, as well as discussing some of the huge range of applications for this versatile technique.

2. FISH: practical considerations

In situ hybridization of genomic DNA probes to chromosomes or nuclei relies on several sequential procedures. The rationale for each of these steps, together with the important parameters, will be considered in turn.

2.1 Preparation of metaphase chromosomes

Metaphase chromosomes can be obtained from any population of actively dividing cells. For gene mapping, metaphase chromosomes are usually prepared by stimulating peripheral blood lymphocytes with phytohaemagglutinin (PHA) and subsequent mitotic arrest using a spindle inhibitor such as colchicine or colcemid. For high-resolution localization of sequences to individual chromosome bands, elongated chromosomes are required. These are achieved by cell synchronization in S phase, followed by release and arrest in early metaphase. It is also advisable to use male metaphase cells for gene mapping, so that all chromosomes (including the Y) are represented. The following protocol (*Protocol 1*) uses a high concentration of thymidine to block the cell cycle. When this block is released, the synchronized population of cells is cultured for a further 5–6 h, and chromosomes harvested after a short exposure to colcemid.

The quality of the chromosome preparations is extremely important to the success of hybridization. Chromosomes should be well spread with little or no cytoplasm present, and should be dark grey when viewed under phase contrast. Proper storage of slides is also important to maintain good quality chromosomal DNA. Slides can be used for hybridization the same day as they are made. However, for long-term storage we keep slides in a sealed container with desiccant at –20 °C.

Protocol 1. Thymidine synchronized blood cultures

Equipment and reagents

- Pre-cleaned microscope slides (Superfrost, BDH)
- 5–10 ml peripheral blood in sodium heparin (20 U/ml) or preservative-free lithium heparin (10 U/ml)
- 15 mg/ml thymidine (Sigma, crystalline)
- 10 μg/ml colcemid (Gibco)

- Lymphocyte culture medium: RPMI 1640, 50 U/ml penicillin, 50 μg/ml streptomycin, 2 mM L-glutamine, 20% fetal calf serum (FCS), 2% PHA (M form) (all from Gibco BRL)
- Hypotonic solution: 0.075 M KCl
- Fixative: 3:1 AnalaR methanol:glacial acetic acid, at 4 °C

Method

1. Add 0.2 ml whole blood to 5 ml of lymphocyte culture medium.
2. Incubate for 48 h or 72 h at 37 °C.[a]

---

3. Add 0.1 ml thymidine (final concentration of 300 µg/ml) and incubate for a further 16–18 h.

4. Centrifuge at 1000 *g* for 5 min. Wash twice in RPMI (no FCS) and resuspend in fresh complete medium (no PHA).

5. Culture for a further 5 h prior to harvesting.

6. 10 min before harvesting add 10 µl colcemid (final concentration 0.02 µg/ml).

7. Centrifuge at 1000 *g* for 5 min. Discard the supernatant and resuspend the pellet in hypotonic solution (pre-warmed to 37 °C). Incubate at 37 °C for 20 min.

8. Centrifuge, then discard the supernatant, and mix the pellet in the small volume of hypotonic solution remaining. Add freshly made fixative dropwise, with mixing. Add the first 1 ml of fixative slowly, then make up to 10 ml.

9. Leave in fixative for 30 min at 4 °C. Centrifuge at 1000 *g* for 5 min, then wash in three to five changes of fixative before making slides.

10. Wipe Superfrost slides clean with absolute ethanol just before use.

11. Place a drop of cell suspension on each slide and air dry. Monitor the quality of chromosome spreading under phase contrast. Chromosomes should be well spread without visible cytoplasm and should appear dark grey under phase contrast (not black and refractile or light grey and almost invisible).

[a] For lymphoblastoid cell lines, grow until a healthy dividing population is obtained, with 0.5–1 × 10⁶ cells/ml (24–48 h after a medium change).

2.2 Probes for FISH

A variety of cloned and uncloned DNA sequences are amenable to FISH procedures. The appropriate selection of primers and amplification of selected DNA by the polymerase chain reaction (PCR) has increased the number and type of DNA which can be used as a probe (reviewed in ref. 7). The large number of ready-labelled commercially available probes has had a considerable impact on clinical cytogenetic analysis. These include:

- repetitive sequence probes (centromeres and telomeres)
- whole chromosome paints
- unique sequence probes

Chromosome-specific centromeric probes target alpha satellite (in some cases beta satellite) DNA families located at chromosome centromeres. Several hundred to several thousand copies of these are present, so that the hybridization signals produced by these probes are extremely strong.

Table 1. Suggested amount of labelled probe and unlabelled competitor DNA required per hybridization (10 μl hybridization mix)

Type of probe[a]	Insert size	Labelled probe DNA	Unlabelled *Cot-1* DNA
Repetitive centromeric[b]	–	10 ng	–
Plasmid[b]	< 1–10 kb	200 ng	–
Bacteriophage λ	Up to 24 kb	200 ng	2.5 μg
Cosmid	35–45 kb	50–100 ng	2.5 μg
Bacteriophage P1	70–100 kb	400 ng–1 μg	5–7.5 μg
PAC	100–300 kb	400 ng–1 μg	5–7.5 μg
YAC (whole yeast DNA)	100 kb–2 Mb	400 ng–1 μg	5–7.5 μg
Alu-PCR YAC DNA	–	200 ng	2 μg
BAC	Up to 300 kb	400 ng	5 μg
Whole chromosome paint (library or PCR-derived)	–	100–400 ng	5–8 μg
DOP-PCR amplified flow-sorted chromosomes	–	100 ng	6.25 μg
Total genomic DNA for CGH	–	200 ng–1 μg	10–50 μg

[a] PAC, P1 artificial chromosome; YAC, yeast artificial chromosome; BAC, bacterial artificial chromosome; DOP-PCR, degenerate oligonucleotide primer-polymerase chain reaction; CGH, comparative genomic hybridization.
[b] No pre-annealing stage.

Centromeric probes are therefore very useful for the detection of chromosome aneuploidy, both in metaphase and interphase cells. Whole chromosome painting probes are complex mixtures of sequences which are used to delineate individual chromosome pairs. These are derived from flow-sorted chromosomes, either as pooled clones from chromosome-specific libraries, or by polymerase chain reaction (PCR) amplification. Whole chromosome paints are now available for all human chromosomes (13, 14). The technique of chromosome painting can be used to identify structural chromosome rearrangements in metaphase (15), although this is of limited use in interphase analysis. Unique sequence probes (genomic or cDNA) cloned in a variety of vectors (see *Table 1*) can be used to target specific chromosome regions, for example for the analysis of microdeletion syndromes and the detection of recurrent translocations in leukaemia.

2.2.1 Preparation of probe DNA

Relatively large quantities of high quality DNA are required for efficient labelling by nick translation. Almost any method for DNA isolation and purification which produces DNA suitable for sequencing will also work for FISH. However, it is worth noting that the most common problems with FISH are probe-related, usually due to probe DNA of poor quality, or due to inaccurate measurement of DNA concentration. The most reliable method for

DNA purification is CsCl gradient centrifugation, but suitable DNA is also obtained by Qiagen column purification (Hybaid), or similar. Minipreps are often not suitable, as they may contain considerable amounts of bacterial DNA. Bacteriophage λ DNA can be prepared by a plate lysis method (16). Labelling of yeast artificial chromosome (YAC) clones contained in crude yeast DNA preparations is often problematic, and large amounts of probe are required for hybridization. Isolation of the YAC from the yeast background by pulsed-field gel electrophoresis (PFGE) is laborious, with a low yield, and in many cases difficult as the YAC may be not visible by ethidium bromide staining. *Alu*-PCR amplification of total yeast DNA will increase the yield of YAC DNA and therefore result in a higher hybridization efficiency (17). However, this technique relies on the spacing and orientation of *Alu* sequences within the YAC, so that *Alu*-poor sequences will not amplify, and the sensitivity for determining YAC chimerism is not known. One method which reproducibly yields total yeast DNA suitable for FISH relies on breaking up the yeast cells by vortexing with small glass beads, followed by phenol:chloroform purification (18).

The other important probe-related factor to consider is DNA concentration. Spectrophotometric measurements are often inaccurate, due to RNA and other contaminants. We prefer to measure probe DNA concentration using a fluorometer (e.g. TKO 100 mini fluorometer, Hoeffer Scientific), which measures the amount of DNA-specific dye which is bound, compared to a known standard. Estimation of DNA concentration on an agarose gel, compared to known standards, is also suitable.

2.2.2 Probe labelling

Probes for localization by FISH are usually labelled with either biotin or digoxigenin, available conjugated to dUTP by a spacer arm of variable length (e.g. bio-16-dUTP, dig-11-dUTP). Deoxyribonucleotides are now also available conjugated directly to fluorochromes including fluorescein isothiocyanate (FITC), tetramethylrhodamine isothiocyanate (TRITC), and aminomethyl coumarin acetic acid (AMCA). Nick translation is the most widely used method for labelling probes for *in situ* hybridization, and is particularly suitable as the fragment size can be controlled by the amount of DNase I in the reaction mixture. This is also an efficient way of labelling large amounts of double-stranded circular DNA, so there is no need to cut out the probe insert before labelling. Indeed, having the vector present means that even short pieces of DNA will be nicked, essential for labelling to proceed. Labelling small uncloned DNA fragments may be problematic as a higher concentration of DNase may be required to ensure that a nick is produced. The size of labelled probe fragments is a critical factor in *in situ* hybridization protocols, with an average size of 300 bp being optimal (range 100–500 bp). Larger probe fragments will result in bright background fluorescence all over the slide, obscuring any specific signal. If the labelled probe fragments are too small (< 50 bp), the site of hybridization may not be visible due to the resulting weak fluorescent signal.

To ensure the correct size of labelled fragments, it is necessary to run a small aliquot of labelled probe on a 2% agarose gel, with *PhiX174/HaeIII* as a molecular weight marker. The DNA should run as a smear between the last six bands of the *PhiX* marker. Other labelling methods (e.g. random primer labelling, PCR) can be used to produce labelled probes for FISH, and may be particularly valuable when the amount of probe available is small. However, in all cases the size of the labelled fragments must be checked, and re-cut with DNase, if necessary.

The protocol given here (*Protocol 2*) uses our own nick translation kit. Commercial kits specifically designed for *in situ* hybridization are also available (e.g. Gibco BRL Bionick kit).

Protocol 2. Labelling of probes by nick translation

Equipment and reagents

- Select B columns (CP Laboratories)
- Purified DNA, whole chromosome library, or PCR amplified chromosomes
- 10 × nick translation buffer: 0.5 M Tris–HCl pH 7.5, 50 mM MgCl₂, 0.5 mg/ml nuclease-free bovine serum albumin (BSA)
- 1 mM biotin-16-dUTP, or 1 mM digoxigenin (dig)-11-dUTP (Boehringer Mannheim)
- 100 mM dithiothreitol (DTT) (Sigma)
- dNTP mix: 0.5 mM each dATP, dCTP, dGTP, and 0.1 mM dTTP (Boehringer)
- DNase I (20 μg/μl) (Amersham International, UK)
- DNase I dilution buffer: 50% glycerol, 0.15 M NaCl, 20 mM sodium acetate pH 5

- 3.5 U/μl DNA polymerase I (Amersham)
- *E. coli* tRNA (10 mg/ml) (Boehringer)
- Salmon sperm DNA (5 mg/ml, sonicated to 200–500 bp) (Sigma)
- TE: 10 mM Tris–HCl pH 7.5, 1 mM EDTA
- Gel loading buffer (5 × bromophenol blue): 10% (w/v) Ficoll, 0.1 M Na₂EDTA, 0.5% (w/v) sodium dodecyl sulfate (SDS), 0.1% (w/v) bromophenol blue
- Electrophoresis buffer (10 × TBE): 108 g Tris base (89 mM), 55 g boric acid (89 mM), 40 ml 0.5 M EDTA pH 8 (2 mM) per litre
- *PhiX174/HaeIII*, size marker (BRL Life Technologies)

Method

1. Add the following (in order) to a 1.5 ml Eppendorf tube on ice:
   - 1 μg probe DNA
   - 1.2 μl 1 mM biotin-16-dUTP or dig-11-dUTP
   - 5 μl dNTP mix
   - 5 μl 10 × nick translation buffer
   - 5 μl 100 mM dithiothrietol (DTT)
   - sterile distilled water to make up to a final volume of 50 μl
   - 3–5 μl 2.5 ng/μl DNase I (need to establish amount for each new batch)
   - 3 μl 3.5 U/μl DNA polymerase I

2. Mix well.

3. Incubate at 15°C for 90 min.

4. Stop reaction by placing tubes on ice.

5. Check the size of the labelled products by running an aliquot on a 2% agarose gel (in TBE and containing 5 μl of 5 mg/ml ethidium bromide per 100 ml) as follows:

    - 5 μl labelled probe (approx. 100 ng)
    - 4 μl gel loading buffer (5 × bromophenol blue)
    - 11 μl sterile distilled water

6. Run at 50 V for 1–1.5 h with *Phi*X174/*Hae*III (20 μl = 250 ng) as a size marker.

7. View on transilluminator and photograph. The optimal size range for *in situ* hybridization is 50–500 bp. A smear of products from 100–300 bp (corresponding to the six smallest bands of *Phi*X174) is suitable. If the size range is larger than this, add a further 5 μl DNase I, place at 15°C for a further 30–60 min, and run another aliquot on a gel to test the size.

8. Purify to remove unincorporated nucleotides by passing the labelled probe through a Select B column (designed for biotinylated probes) according to the manufacturer's instructions.

9. Measure the volume of eluate then ethanol precipitate the purified, labelled probe by adding:

    - 50 μg *E. coli* tRNA
    - 50 μg salmon sperm DNA
    - 0.1 vol. 3 M sodium acetate pH 5.6
    - 2–2.25 vol. ice-cold ethanol

    Mix well and place at –70°C for 1–2 h or –20°C overnight.

10. Centrifuge in a microcentrifuge for 15–25 min at 4°C. Pour off the supernatant and dry the pellet (either air dry or in a vacuum desiccator). Resuspend the pellet in 20 μl TE pH 8 to give a final concentration of 50 ng/μl. Allow the DNA to dissolve at room temperature for 1–2 h or at 4°C overnight with occasional mixing. Purified, labelled probes are stable for several years when stored at –20°C.

2.3 Pre-treatment of chromosomal DNA

The methanol:acetic acid fixation of metaphase chromosomes removes some basic proteins which might interfere with hybridization. However, there is still a variable amount of other protein and cytoplasmic contaminants on metaphase chromosome preparations which may block hybridization, or cause non-specific background. We routinely use an RNase pre-treatment to reduce background due to RNA:DNA hybridization, in combination with formaldehyde fixation steps to preserve chromosome morphology. For some chromosomal targets (e.g. bone marrow smears, bone marrow progenitors from methyl cellulose-grown colony assays, tumour preparations), it is necessary to add a proteolytic digestion treatment to this. In this case we use pepsin (0.1% in 0.01 M HCl, freshly made) for 7 min at 37°C, after the RNase step. However, over-digestion can also cause problems (loss of cells from the slide), so only use when absolutely necessary.

Protocol 3.   Pre-treatment of slides

Equipment and reagents

- Slide containing metaphase chromosomes
- PBS/50 mM MgCl$_2$: 50 ml 1 M MgCl$_2$ plus 950 ml phosphate-buffered saline (PBS)
- PBS/50 mM MgCl$_2$/1% formaldehyde (make up fresh each time): 2.7 ml formaldehyde in 100 ml PBS/MgCl$_2$

- PBS: 8 g NaCl, 0.2 g KCl, 1.15 g Na$_2$HPO$_4$, 0.24 g KH$_2$PO$_4$ in 800 ml H$_2$O, pH to 7.4 with HCl; add H$_2$O to 1 litre
- RNase A (10 mg/ml) (Sigma): boiled for 10 min to remove contaminating DNase
- 40% formaldehyde (w/v)

Method

1. Place 100 μl RNase (100 μg/ml) on slides under a 24 × 50 mm coverslip and incubate at 37°C for 30 min–1 h.

2. Wash twice (3 min each) in 2 × SSC (with agitation).

3. Place slides in PBS/50 mM MgCl$_2$ for 5 min.

4. Fix in PBS/50 mM MgCl$_2$/1% formaldehyde for 10 min.

5. Wash in PBS for 5 min (with agitation).

6. Dehydrate slides through an alcohol series (70%, 95%, 100%) and allow to air dry. Slides can be stored desiccated at 4°C for up to one month before use.

2.4 Competitive *in situ* suppression hybridization (19, 20)

Clones containing large DNA fragments (i.e. phage, cosmid, YAC, P1) and whole chromosome paints require an additional step before hybridization to remove ubiquitous repetitive sequences (see *Table 1*). This is achieved by a short incubation prior to hybridization, with unlabelled human competitor DNA, in the form of either total human DNA (placental DNA, sheared and sonicated to 50–300 bp) or human *Cot-1* DNA (Gibco BRL). We prefer *Cot-1* DNA for most purposes, as this requires a shorter incubation time (15 min). However, there may be occasions when moderately repeated sequences are not blocked by *Cot-1*. In these cases, total human DNA (such as placental DNA) may be more suitable. It is necessary to use approximately twice as much as *Cot-1* DNA, and may even be necessary to titrate the amount of total human DNA to determine the correct amount of competitor. Pre-hybridization annealing is performed for 1–3 h with total human DNA. Repetitive sequence probes and single copy clones containing small (< 3 kb) inserts, such as plasmids and cDNAs, usually do not require pre-hybridization annealing with competitor.

The principles of *in situ* hybridization to chromosome preparations are the same as for filter hybridization, but need to take into account the nature of the target DNA (i.e. chromosomal DNA on microscope slides). This restrains the choice of reagents and temperature for some steps: for example, denaturing with harsh alkali or at high temperatures will cause cells to be completely lost from the slide. The important parameters to consider are:

(a) The hybridization temperature. The temperature at which two DNA strands separate (T_m) is in the range 85–95 °C. The optimal DNA:DNA reassociation temperature (T_r) is approx. 25 °C below the T_m of the native duplex. However, fixed chromosome preparations on microscope slides will not tolerate temperatures > 65 °C for long periods of time. The presence of formamide in the hybridization buffer lowers the T_r, allowing hybridization to take place at 37–42 °C, thus preserving chromosome morphology.

(b) Time of hybridization. This depends on the size and copy number of the target sequence, as well as the complexity of the probe. Repetitive sequence probes such as alphoid centromeric probes require only 1 h for hybridization. Unique sequence probes cloned in plasmid, cosmid, or phage vectors require hybridization overnight (16–18 h). Larger insert probes (large YACs) may benefit from longer (one to two days), and very complex probes such as whole genomes in comparative genomic hybridization (CGH) require two to three days.

(c) Denaturation of chromosomal DNA. The optimal time for denaturation needs to be determined for each batch of slides. Over-denaturation results in loss of chromosome morphology and very poor DAPI banding after hybridization. Under-denaturation will result in little or no signal. Most protocols err on the side of over-denaturation: denaturation solutions containing 70% formamide should need only 2–3 min at 72 °C, if slides are pre-warmed on a hot plate at 50–60 °C. The pH of the denaturation solution is also important: this should be checked when the solution is at the desired temperature and adjusted if necessary. It is preferable to prepare the denaturing solution and use as soon as it has reached the desired temperature to prevent pH fluctuations. We use EDTA (final concentration 0.1 mM) to stabilize the denaturing solution against pH changes.

(d) Stringency of hybridization conditions. Renaturation depends on specific base pairing between two complementary DNA strands, and can be controlled by the stringency of the hybridization conditions. Increasing the hybridization temperature or decreasing the salt concentration will increase the stringency, which has a direct effect on the accuracy of base pairing.

Protocol 4. Competitive *in situ* suppression hybridization[a]

Reagents

- Human *Cot-1* DNA (BRL Life Technologies)
- 3 M sodium acetate
- Denaturing solution: 70% (v/v) formamide, 2 × SSC, 0.1 mM EDTA pH 7
- Hybridization buffer: 50% (v/v) formamide, 10% (w/v) dextran sulfate, 1% (v/v) Triton X-100, 2 × SSC pH 7
- Formamide (Fluka)

- 50% dextran sulfate
- 20 × SSC: 3 M sodium chloride, 0.3 M sodium citrate pH 7
- Blocking solution: 3% (w/v) BSA in 4 × SSC, 0.05% (v/v) Triton X-100 (make up fresh)
- Wash solution: 4 × SSC, 0.05% (v/v) Triton X-100

Method

1. Dry down the appropriate concentration of probe and competitor (see *Table 1*) either in a vacuum desiccator (SpeedVac) or by ethanol precipitation; e.g. for cosmids:

   - 100 ng labelled probe (usually 2 µl)
   - 2.5 µg (2.5 µl) *Cot-1* DNA
   - 0.1 vol. 3 M sodium acetate
   - 2 vol. ice-cold ethanol

   Allow to precipitate for 1–2 h at –70C.

2. Centrifuge and dry down the pellet as for labelled probes. Resuspend pellet in 11 µl hybridization buffer (warmed to room temperature).

3. Denature the probe mixture at 95°C in a hot block for 10 min. Plunge the tubes on ice for a few minutes, then centrifuge briefly in a micro-centrifuge.

4. Place the probe mixture at 37°C for 15 min–2 h.

5. Just prior to hybridization, denature the chromosomal DNA as follows:

   (a) Incubate slides in denaturing solution (in water-bath in a fume-hood) at 70°C for 5 min.

   (b) Wash slides in cold 2 × SSC, followed by two changes of 2 × SSC.

   (c) Dehydrate through a cold alcohol series (70%, 90%, 100%).

6. Air dry the slides and place on a hot plate at approx. 42°C.

7. Centrifuge the probe mixture quickly to get the liquid to bottom of tube. Place this mixture on the previously treated slide containing chromosomes and cover with 22 × 32 mm coverslip (do not let drop dry). Seal the coverslip with rubber solution and place the slides in a moist chamber at 37°C for overnight–four days.

8. Remove rubber solution. Coverslips can then be removed either by soaking in 2 × SSC or gently tipping them off into the glass disposal bin (never pull them off!).

9. Carry out the following washes:[b]

   (a) Three washes (3 min each) in 2 × SSC at room temperature (with agitation).

   (b) Two washes (20 min each) in 0.1 × SSC at 65°C.

   (c) One 5 min wash in 0.1 × SSC at room temperature (with agitation).

10. Wash slides in wash solution for 3 min.

11. Incubate slides in blocking solution for 10–20 min (room temperature).

12. Wash in wash solution for 3 min before carrying out the appropriate detection steps.

[a] With alphoid repetitive probes and unique small insert clones (< 1–3 kb), the whole probe contributes to the hybridization signal. The appropriate amount of labelled probe (see *Table 1*) is dried and resuspended in 10 μl hybridization buffer. Denature the probe mixture at 75°C for 5 min, then place on ice for 5 min. Follow *Protocol 4*, step 5 as usual.
[b] We have found that these washes give the same result as standard formamide washes, i.e. 50% formamide, 2 × SSC (15 min), 2 × SSC (15 min) at 42°C.

2.5 Detection of bound, labelled probe

The development of monoclonal antibody technology, the efficient chemical conjugation of reporter molecules, the use of the biotin–avidin affinity system, and the availability of sensitive fluorescence imaging systems have all contributed to the wider use of FISH. Fluorescent detection of biotin labelled probes uses sequential layers of avidin, biotinylated anti-avidin antibodies, and avidin–fluorochrome. Digoxigenin labelled probes are detected by specific antisera followed by fluorescently-conjugated anti-immunoglobulins. Fluorochromes available for FISH are given in *Table 2*. The detection of single probes labelled with either biotin or digoxigenin usually uses FITC, with chromosomes counterstained with PI and DAPI. The mixture of DAPI and PI gives an R-banded pattern when viewed under the green filter set and a G-banded pattern under the ultraviolet filter set. However, high-performance

Table 2. Fluorochromes available for FISH

Fluorochrome[a]	Wavelength range (nm)	Colour
AMCA	425–475	Blue
FITC,	500–550	Green
TRITC, Cy3	550–600	Orange
Texas red	600–650	Scarlet
Cy5	650–700	Far red

[a] TRITC, tetramethylrhodamine isothiocyanate; FITC, fluorescein isothiocyanate; AMCA, aminomethyl-coumarin acetic acid.

cooled CCD cameras are too sensitive to allow PI as counterstain, and DAPI is preferred. In this case, it may be preferable to detect singly-labelled probes with Texas red to give more contrast against blue chromosomes. Dual colour detection of two different probes is carried out in a mixture of biotin–Texas red (red signal) and anti-digoxigenin–FITC (green signal). To obtain an even intensity for both probes, label the weaker probe with biotin.

Protocol 5. Detection of bound, labelled probe

Equipment and reagents

- Fluorescence microscope (epifluorescence illumination), with suitable fluorescence objectives and filter sets (usually need separate filter sets for FITC, Texas red/rhodamine, and DAPI/AMCA, as well as a double or triple filter block)
- Moist chamber for antibody detection steps: use a plastic microscope slide box (Raymond Lamb) containing moist tissue paper (wring out excess water), placed in incubator or floated in water-bath. Alternatively, use metal trays (Lamb's immuno slide staining trays, Raymond Lamb) for both hybridization and detection steps.
- Avidin DCS–FITC (1 mg/ml) (Vector, Cat. No. A-2011)
- Biotinylated anti-avidin D (0.5 mg/ml) (Vector, Cat. No. BA-0300)

- Propidium iodide (Sigma)
- 4′,6-diamidino-2-phenylindole (DAPI) (Sigma)
- Vectashield mountant (Vector Labs)
- Avidin DCS–Texas red (2.5 mg/ml stock) (Vector Labs)
- Diluent for antibodies: blocking solution (see *Protocol 4*), filtered through a 0.45 μm syringe filter—stock antibody solutions are stored at –20°C
- Monoclonal anti-digoxigenin (Sigma, Cat. No. D8156)
- Rabbit anti-mouse Ig–FITC (Sigma, Cat. No. F9137)
- Monoclonal anti-rabbit–FITC (Sigma, Cat. No. F4890)

A. *Biotin labelled probes*

1. Dilute 2.5 μl of stock avidin DCS–FITC in 1 ml blocking solution (5 μg/ml final concentration). Use 100 μl of this under a 24 × 50 mm coverslip. Incubate in a moist chamber at 37°C for 30 min.

2. Flick off coverslips and wash slides three times (for 3 min each) in wash solution (see *Protocol 4*).

3. Dilute 10 μl of stock biotin anti-avidin D in 1 ml blocking solution (5 μg/ml final concentration). Use 100 μl of this under 24 × 50 mm coverslip. Incubate in a moist chamber at 37°C for 30 min.

4. Flick off coverslips and wash slides three times (for 3 min each) in wash solution.

5. Add 100 μl avidin–FITC (same as the first layer). Incubate for 30 min under coverslip as before.

6. Carry out the following final washes:

   (a) Wash once for 3 min in wash solution.

   (b) Wash twice (5 min each) in PBS.

   Dehydrate slides through an ethanol series. Air dry.

---

7. Mount slides in 40 μl Vectashield containing 1.5 μg/ml 4',6-diamidino-2-phenylindole (DAPI) and 0.75 μg/ml propidium iodide (PI) under a 24 × 50 mm coverslip. Seal the edges of the coverslip with rubber solution or nail varnish. The signal keeps well for several weeks if slides are stored at 4°C.

B. *Dixogeninin labelled probes*

1. Prepare all antibody dilutions in blocking solution, filtered before use. Make up the following antibody dilutions in 1 ml blocking solution:

   (a) 1st layer: 1.5 μl mouse monoclonal anti-digoxigenin.

   (b) 2nd layer: 1 μl rabbit anti-mouse–FITC.

   (c) 3rd layer: 10 μl monoclonal anti-rabbit–FITC.

2. Incubate sequentially in each antibody layer (100 μl under a 24 × 50 mm coverslip) for 30 min at 37°C in a moist chamber. After each antibody layer, wash three times (3 min each) in wash solution.

3. Carry out final washes as for biotin detection (*Protocol 5A*, step 6).

4. Mount in Vectashield containing 1.5 μg/ml DAPI and 0.75 μg/ml propidium iodide.

C. *Dual colour detection of biotin and digoxigenin labelled probes*

1. Prepare all antibody dilutions in blocking solution, filtered before use. Make up the following antibody dilutions in 1 ml blocking solution:

   (a) 1st layer: 1 μl avidin–Texas red plus 1.5 μl mouse monoclonal anti-digoxigenin.

   (b) 2nd layer: 10 μl biotin–anti-avidin plus 1 μl rabbit anti-mouse–FITC.

   (c) 3rd layer: 1 μl avidin–Texas red plus 10 μl monoclonal anti-rabbit–FITC.

2. Incubate in each antibody layer for 30 min at 37°C in a moist chamber. After each antibody layer, wash three times (3 min each) in wash solution.

3. Carry out final washes as for biotin detection (*Protocol 5A*, step 6).

4. Mount in Vectashield containing 1.5 μg/ml DAPI only.

2.5.1 Multicolour hybridization

The simplest method for the simultaneous hybridization and detection of three different probes can be achieved using one probe labelled with biotin (detected with Texas red, giving a red signal), a second probe labelled with digoxigenin (detected with FITC, giving a green signal), and a third probe labelled separately with both biotin and digoxigenin, mixed in a 1:1 ratio (detected in dual colour reagents, resulting in a yellow signal). To extend this

Table 3. Mixing ratio for simultaneous detection of six targets

Probe	TRITC[a]	FITC[a]	Biotin–AMCA[a]	Colour
1	1	–	–	Red
2	–	1	–	Green
3			1	Dark blue
4	1	1	–	Yellow
5	–	1	1	Blue-green (aqua)
6	1	–	1	Red-blue (purple)

[a] TRITC, tetramethylrhodamine isothiocyanate; FITC, fluorescein isothiocyanate; AMCA, aminomethyl-coumarin acetic acid.

approach for the detection of six different probes, three different haptens are required for labelling (21). This can be achieved using probes directly labelled with FITC–dUTP, TRITC–dUTP, and indirectly with biotin, detected with avidin–AMCA (22). The latter approach is most suitable for repetitive sequence probes and whole chromosome paints, but has been reported for single copy cosmid and plasmid probes, when used in conjunction with CCD imaging. Probes are labelled in different nick translation reactions, then mixed in the final hybridization solution (*Table 3*).

2.6 Visualization of signal

Fluorescent signal from all except the smallest probes can be seen using any modern epifluorescence microscope equipped with the appropriate filter sets. However, obtaining a permanent record of these relatively weak fluorescence images presents some difficulties. Conventional photomicroscopy of multi-colour FISH images is difficult, principally because of the long exposure times and loss of registration of images when changing filters. Double or triple exposures, fast film, and a perfectly aligned microscope are prerequisites. Perfect registration of fluorescent signal to chromosome bands is essential for gene mapping applications. Even on the best aligned microscopes, changing fluorescent filters results in some image shift. This problem can be overcome in several ways. Dual and triple bandpass filter sets are available so that several fluorochromes can be observed without changing filters. However, the emission of all fluorochromes is significantly reduced, which may make it difficult to see weak signals. Sensitive digital imaging systems such as confocal laser scanning microscopes and cooled charge-coupled device (CCD) cameras have overcome the problems of image registration and provide other significant advantages over conventional fluorescence microscopy. Confocal laser scanning microscopes were designed primarily to provide thin sections through relatively thick tissue by light microscopy, but also provide significant advantages for imaging FISH signals. They provide complete and accurate registration of fluorescent signals on chromosomes by the simultaneous scanning of each fluorochrome through separate filter blocks, followed by merging and pseudocolouring of the two monochrome images. Weak fluorescent signals can be amplified, and the resultant images stored into computer memory and subjected to further image processing.

The introduction of cooled CCD cameras to fluorescence microscopy has increased both the sensitivity and flexibility of FISH techniques. High-performance, highly cooled (–30 °C) CCD cameras are several orders of magnitude more sensitive than confocal laser scanning devices and can detect fluorescence signals that are not visible by the human eye. However, video-rated, ambient temperature (+15 °C) CCD cameras are probably sufficient for most FISH applications, and some applications (i.e. chromosome painting) only require relatively inexpensive video cameras. All of these imaging systems provide a significant advantage over confocal microscopy in terms of multicolour imaging, as the number of fluorochromes which can be detected depends only on the number of filters that can be fitted on the microscope. Problems with image registration due to the movement of microscope filter blocks can be overcome by the use of a filter wheel containing the excitation filters and situated between the lamp and the microscope. Most CCD imaging systems also come with software for chromosome analysis and CGH, in addition to FISH image capture and enhancement facilities.

2.6.1 Interpretation of results

FISH provides a direct way of assigning chromosomal location for cloned probes and is also the best way to detect chimerism of large clones such as YACs. Metaphase chromosomes can be used to resolve probes 2–3 Mb apart. Mechanically stretched chromosomes can give information on the order of probes > 200 kb apart, and is useful to determine which probes are more centromeric or telomeric (23, 24). For probes closer than this, the relatively decondensed chromatin in interphase nuclei can be used. Interphase mapping can resolve probes 50–500 kb apart, and can determine relative probe order, but gives no information on chromosome location, or telomere–centromere orientation. To increase the resolution of FISH mapping even further, several new procedures use extended DNA fibres, allowing the resolution of probes 1 kb apart. The resolution of different FISH mapping techniques is given in *Table 4*.

Hybridization efficiency decreases proportionately with decreasing target size. For repetitive sequences such as centromeric alphoid repeats, hybridization signal is relatively large, irrespective of the size of the sequence, because of the high copy number. Large insert probes such as cosmids, YACs, BACs, P1, PACs hybridize very efficiently (> 80% of cells with signal on all four chromatids), so that usually only a few cells (five to ten) need to be scored. Small single copy sequences (< 3 kb) hybridize less efficiently (30% of cells with signal on all four chromatids), and many more metaphases need to be evaluated (up to 30). In extreme cases, for example with cDNA clones

Table 4. FISH mapping techniques

Target	Resolution	Application	Ref
Metaphase, prometaphase chromosomes	1–3 Mb	Chromosome and band assignment, detection of clone chimerism.	–
Mechanically stretched chromosomes	> 200 kb	Clone ordering. Useful for centromere–telomere orientation.	23, 24
Interphase nuclei	50 kb–1 Mb	Clone ordering and orientation. Centromere–telomere orientation not possible.	8, 30
DNA fibres	1–500 kb	Clone ordering, orientation. Assessment of overlaps, gaps in contig map.	10, 31–36
Molecular combing	1–500 kb	Accurate distance measurements, e.g. precise localization of clones along YAC.	11, 12

(< 1 kb insert), it may be necessary to count the signals and plot an ideogram to obtain a confident localization.

3. Troubleshooting

Frequently encountered problems:

(a) No hybridization signal. This may be due to:

    (i) Insufficient probe DNA in the hybridization mix. The DNA concentration of any new probe should be measured accurately on a fluorometer, or on an agarose gel including ethidium bromide against a range of concentrations of uncut λ DNA.

    (ii) Inadequate denaturation of probe and/or chromosomes.

    (iii) Probe fragment size too small. Always check the labelled fragment size on a 2% (1.2% for PCR products) agarose gel with *Phi*X174/*Hae*III size marker. The optimum fragment size is 100–500 bp.

(b) High background. High background with strong specific signal may be due to:

    (i) Low stringency of hybridization or post-hybridization washes.

    (ii) Incomplete competition.

The stringency of hybridization can be increased by either increasing the hybridization temperature, increasing the formamide concentration of the hybridization mix and/or post-hybridization washes to 60%, or decreasing the SSC concentration to 0.1× in the post-hybridization washes. Alternatively, increase the *Cot-1* DNA concentration: this is already present in large excess so that any increase should be substantial (up to tenfold).

(c) Brightly fluorescent signal all over the slide. This occurs when the labelled probe fragments are too large: if labelled probe is > 500 bp it should be re-cut with DNase I. High background of this type may also be caused by insufficient blocking with BSA.

(d) Cells lost from slide. Handle slides with care at all stages especially during removal of coverslips (never pull them off!). Agitation during post-hybridization washes should be carried out on a rocking platform set at minimum speed.

(e) Poor chromosome morphology/banding. If chromosomes look puffy they may have been over-denatured: always check temperature of denaturing solution inside the Coplin jar. Over-denatured chromosomes give a C-banding pattern with DAPI staining.

4. Applications

4.1 Gene mapping

4.1.1 Localization of sequences to chromosome bands

The accurate localization of cloned sequences to chromosome bands by FISH can be achieved using a variety of methods. Fractional length measurements can be used with some accuracy (25), but this may be difficult to relate to standard cytogenetic nomenclature (26). Other banding methods which can be used include:

(a) Incorporation of BUdR into chromosomal DNA during harvesting followed by staining with Hoescht 33258 and UV irradiation, resulting in an R-banding pattern (27).

(b) Counterstaining with a mixture of DAPI and PI (28). This results in a Q-banding pattern when viewed through the UV filter set, and an R-banding pattern when viewed through the green filter set.

(c) Concurrent banding with *Alu*-PCR products labelled with a fluorochrome (29).

We find that DAPI banding using (b) gives satisfactory results in most cases. Most FISH imaging systems can produce an enhanced G-banded image from the weaker fluorescent banding pattern produced by DAPI staining which is extremely useful for gene mapping (*Figure 1a*).

4.1.2 Interphase mapping

The relationship between interphase distance and physical distance is linear in the range 50–1000 kb, but is more reliable in the range 50–500 kb (8, 30). Interphase mapping by FISH should preferably be carried out on cells in G1, where the hybridization signal is seen as a single dot on each chromosome. To determine relative order, at least three probes are required (two with known

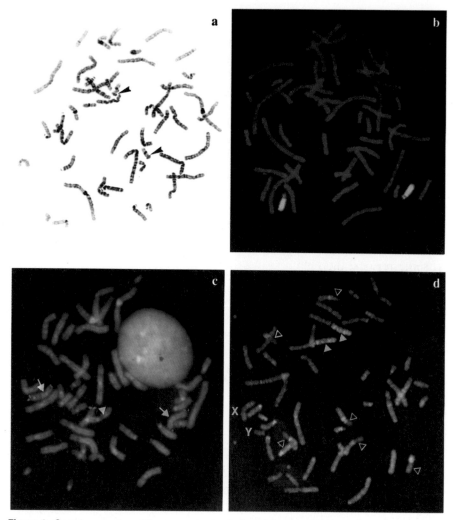

Figure 1. Some examples of fluorescence *in situ* hybridization (FISH). (a) Fine mapping of a biotin labelled cosmid probe to 19q13.3 (arrowheads). The Texas red signal corresponding to the probe is overlaid onto a G-banded image obtained from DAPI stained chromosomes using *MacProbe* software (PSI, Chester, UK). (b) Reverse chromosome painting of a probe derived by DOP-PCR amplification of flow-sorted abnormal chromosomes 18 to a normal male metaphase. The chromosomes are counterstained with propidium iodide (red). Fluorescent signal (yellow) corresponding to the fluorescein detected paint highlights only part of the chromosome 18p arm, confirming that this is a del(18p). (c) Dual colour FISH with probes specific for the 10p and 10q subtelomeric regions demonstrates the presence of a translocation of 10q subtelomeric sequences to 16p (arrowhead). The 10p probe is labelled in biotin and detected in Texas red (red), and the 10q probe labelled in digoxigenin and detected with FITC (green). The DAPI counterstained chromosomes appear light blue. The two normal chromosome 10 homologues are shown by arrows. (d) Comparative genomic hybridization (CGH) using tumour DNA from the breast cancer cell line MPE 600 demonstrates amplification of 1q (green) (closed arrowheads), and deletion of 16q, 9p, 11q (open arrowheads). The Y chromosome appears red and the X chromosome green, as a result of hybridizing female tumour and control DNA to normal male metaphases.

order), differentially labelled and detected in two different colours. Metaphase chromosome preparations can be used for interphase mapping as they contain large numbers of interphase cells. However, a considerable proportion of these are usually in G2 with the resultant FISH hybridization signals appearing as doublets, making interpretation difficult. Interphase nuclei in G1 can be prepared from fibroblast cell lines which are grown to confluency, with all nuclei the same size, and no mitotic cells seen. Analysis of 20–30 cells is usually sufficient to give a reliable order.

4.1.3 Extended DNA preparations (Fibre-FISH)

A large number of techniques under the generic term Fibre-FISH have been employed to provide free, linearly extended DNA and thereby increase the resolution of FISH mapping (see *Table 4*). Halo preparations use DNA released by detergent treatment to form loops around the nuclear matrix (9). In some regions, the halos become detached and allow a linear analysis of probes. Hybridized cosmid probes are detected as a string of fluorescent signals. This method has the advantage of allowing the discrimination of both chromosome homologues, and may be useful for colour 'bar coding' of deletions of large genes (31, 32). The DIRVISH technique also uses detergent to provide a linear stream of DNA along the length of a microscope slide (33). Fibre-FISH techniques can now resolve sequences < 1 kb apart, and have been used to localize exons and cDNAs within cosmids (34). They also permit hybridization over large distances, allowing the ordering and orientation of fragments from within cosmids and YACs (35). Another novel application for this technique is the demonstration of translocations where the breakpoints are dispersed over a large distance (36). However, the accuracy of distance measurements using any of the above techniques is debatable. The technique of molecular combing claims to produce DNA strands which are reproducibly stretched, and can be used to estimate distances between clones and the amount of overlap (11, 12). These appear to be most reliable if used on a subset of DNA (e.g. purified YAC DNA), rather than the whole genome.

4.2 Identification of genetic material of unknown origin

The challenge of identifying additional chromosomal material with no clues as to the origin (e.g. marker chromosomes, *de novo* unbalanced translocations) can be approached in several ways. Reverse chromosome painting may be informative, provided there is a cell line available with the abnormal chromosome, and facilities for chromosome sorting. However, degenerate oligonucleotide primer-PCR amplified chromosome paints do not cover the whole genome evenly, and the sensitivity of this technique for the detection of subtelomeric chromosome rearrrangements is unknown. Microdissection of small marker chromosomes of amplified segments is an elegant approach, particularly for the analysis of small marker chromosomes which are not separable by flow sorting, but requires specialized technology. Comparative genomic hybridization (CGH) is the method of choice for solid tumours and certain types of leukaemia, where it may be difficult to obtain metaphase chromosomes. Finally, 24 colour chromosome painting will identify all chromosome pairs in a single hybridization. This is certainly the way of the future, but awaits further developments to make the technique widely accessible.

4.2.1 Reverse chromosome painting

In this approach (*Protocol 6*), the abnormal chromosome fraction is purified by fluorescence-activated cell sorting and amplified using PCR and a degenerate oligonucleotide primer (DOP-PCR). When the amplified product is used as a probe for FISH to normal metaphase chromosomes, the origin of the abnormal chromosome is revealed (37, 38) (see *Figure 1b*). We have used this approach to identify the origins of three *de novo* abnormal chromosomes 16 (39). In another study of 99 patients with unexplained mental retardation, we used reverse painting to show that two out of three cases of apparent deletion were unbalanced translocations (40).

Protocol 6. DOP-PCR amplification of flow-sorted normal and abnormal chromosomes

Reagents

- Flow-sorted chromosomes (approx. concentration 500/µl)
- 2 × PCR buffer: 10 mM MgCl$_2$, 100 mM KCl, 20 mM Tris–HCl pH 8.4, 0.2 mg/ml gelatin
- dNTP mix: 2 mM each dATP, dCTP, dGTP, dTTP
- 6-MW primer: 5′ CCGACTCGAGNNNNNNN-ATGTGG 3′ (30 µM)
- 2.5 U/µl *Taq* 1 polymerase (Boehringer Mannheim)
- 1 mM biotin-16-dUTP or 1 mM digoxigenin-11-dUTP (Boehringer Mannheim)

Method

1. Combine in a sterile 0.5 ml microcentrifuge tube:[a,b,c]

   - x µl (= 500 flow sorted chromosomes)
   - 50 µl 2 × PCR buffer
   - 10 µl dNTP mix
   - 6.6 µl 30 µM 6-MW primer
   - 0.5 µl (= 1.25 U) *Taq* 1 polymerase
   - water to a final volume 100 µl

2. Overlay with 100 µl mineral oil and run the following program in a DNA thermal cycler:

   (a) Denature for 10 min at 93°C.

   (b) Five cycles of: 1 min at 94°C, 1.5 min at 30°C, 3 min at 30–72°C transition, 3 min at 72°C.

   (c) 35 cycles of: 1 min at 94°C, 1 min at 62°C, 3 min at 72°C , with an additional 1 sec/cycle, and final extension time of 10 min.

3. Run a 10 μl aliquot of the amplified products on a 1.2 % agarose gel with *Phi*X174 to check the success of the amplification. There should be no amplification in the negative control.

4. For second round PCR and labelling, add to a new sterile 0.5 ml microcentrifuge tube:
   - 5 μl of amplified products from first round
   - 25 μl 2 × PCR buffer
   - 5 μl nucleotide mix
   - 3.3 μl 6-MW primer
   - 12 μl 1 mM biotin-16-dUTP
   - 0.25 μl *Taq*I polymerase

5. Mix well, overlay with 50 μl mineral oil, and run the following PCR program:
   (a) Denature for 10 min at 93 °C.
   (b) 25 cycles of: 1 min at 94 °C, 1 min at 62 °C, 3 min at 72 °C, with a final extension time of 10 min.

   Remove the mineral oil. Run 10 μl of labelled products on a 1.2% agarose gel to check the size range. If the labelled fragments are too large, re-cut with 5 μl DNase I for 30–60 min.

6. Purify the labelled DNA through a Select B column (see *Protocol 1*). Measure the DNA concentration of the purified, labelled DNA in a fluorometer (usually 20–50 ng/ml). Ethanol precipitate the labelled DNA with tRNA and salmon sperm DNA as usual, dry, and resuspend in H$_2$O or TE to a suitable concentration: this is now ready for use as a chromosome paint. Use 100 ng probe plus 6 μg *Cot-1* DNA per slide.

[a] All of these reagents except chromosomal DNA and *Taq* 1 polymerase can be sterilized by exposure to short wave UV irradiation (5 min on a transilluminator).
[b] All solutions, microcentrifuge tubes, and tips should be autoclaved and kept for PCR only. Use aerosol-resistant tips and add all reagents in laminar flow-hood to minimize contamination.
[c] Prepare positive (2.5 pg genomic DNA) and negative (all of the reagents except chromosomes) controls in the same way.

4.2.2 Chromosome microdissection

This powerful technology allows the generation of region-specific probes for any part of the human genome. Hybridization probes are prepared by microdissection of human chromosomes followed by DOP-PCR amplification and labelling of the microdissected DNA fragments (41, 42). This technique is invaluable in the identification of intractable marker chromosomes such as homogeneously staining regions and in the accurate identification of chromosomal breakpoints (43, 44).

4.2.3 Multicolour chromosome painting

Multicolour FISH using the method of 'combinatorial' probe labelling was first described by Nederlof *et al.* (45) using chromosome-specific centromeric probes. Since then, many groups have worked towards an approach which would allow the simultaneous hybridization of 24 different painting probes to a single metaphase to identify all 24 pairs of human chromosomes (21, 46). Theoretically, this is possible using only five different fluorochromes. The difficulty in achieving the discrimination of 24 different colours in this way lies in the limited number of fluorochromes which can be spectrally separated.

However, two recent reports seem to make this an attainable goal (5, 6). The first group were able to visualize 27 different colours on a single metaphase using combinatorial labelling with five different haptens or fluorochromes (5). This was achieved by careful designing of specific filter sets in combination with sophisticated, dedicated software. The second elegant approach to the same problem also employed combinatorial labelling with only five fluorochromes (6). In this technique, called spectral karyotyping, the discrimination of differentially labelled chromosomes was based on a measurement of the spectral characteristics of each mixture of colours, using a combination of Fourier spectroscopy and cooled CCD imaging. This does not rely on selecting specific filter sets, as the spectral characteristics of each chromosome are determined on a pixel-by-pixel basis, and compared with computed ratios for a particular chromosome. Multicolour chromosome painting now looks set to revolutionize the analysis of complex karyotypes, although the more widespread use of these techniques will depend on the further development of both instrumentation and ready-labelled probes.

4.3 Comparative genomic hybridization (CGH)

This technique provides a way of identifying new regions of amplification or deletion in many different types of malignancy without the need to obtain dividing cells. Normal and test (tumour) DNA are differentially labelled, mixed together, and the resultant mixture used as a complex probe to normal metaphase chromosomes (47). Detection of the two genomes with two different fluorochromes allows the visualization of copy number difference along the length of the chromosome. The fluorescent signals are quantitated using a digital image analysis system, with a software program to calculate the ratio of green to red fluorescence along the length of each chromosome. Gross differences (gain or loss of a whole chromosome) can be seen through the triple bandpass filter of a fluorescence microscope (see *Figure 1d*). However, in order to accurately detect small differences, quantitative measurements of the fluorescence ratios is required, with statistical analysis of the resultant ratios. Apart from the technical difficulties in establishing this technique, there are several important limitations. Firstly, it will not detect balanced rearrangements, such as translocations or inversions. Secondly, reliable detection of amplifications or deletions requires the presence of at least 50% abnormal cells. It is estimated that amplifications of > 2 Mb and deletions of 10–20 Mb can be detected. The limiting factor in this sensitivity is the resolution of metaphase chromosomes, as it is estimated that the smallest region that can be adequately discriminated by FISH is approximately 3–5 Mb (48). Despite these reservations, CGH has proved a robust and accurate technique, which provides access to tumours not amenable to conventional cytogenetic analysis (49, 50). This analysis is the first step to identifying new regions containing genes important in oncogenesis.

4.4 Breakpoint identification

In combination with intensive gene mapping strategies, FISH provides a powerful tool for the analysis of chromosomal breakpoints in genetic disease and cancer. The isolation of a series of YAC or cosmid clones covering a particular disease region (for example the chromosomal breakpoints of consistent translocations in leukaemia), provides the starting point. FISH with these sequences to leukaemic metaphase chromosomes can rapidly and easily identify the clone containing the breakpoint, as the fluorescent signal will be split over the two derivative chromosomes involved in the translocation. Signal on the normal chromosome homologue provides an internal control. FISH can also be used for determining the minimal deleted region in non-random deletions in leukaemia, thus narrowing the regions containing tumour suppressor genes. In this way, FISH has significantly speeded up the process of identifying genes implicated in oncogenesis, and associated with specific genetic diseases.

4.5 Interphase analysis

The ability to analyse chromosomal abnormalities in interphase cells by FISH provides significant advantages over metaphase FISH in a variety of situations, although only limited information can be gained by this approach. In oncology, the main benefits are the ability to screen large number of cells, the lack of a need for culture, and access to a variety of cell types (51, 52). For example, trisomy 12 is the most common clonal cytogenetic abnormality detected in chronic lymphocytic leukaemia (CLL), but conventional cytogenetic analysis is unsuccessful or yields only normal karyotypes in up to 50% of cases. Interphase FISH studies of CLL have now demonstrated a higher frequency of trisomy 12 than determined by G-banding analysis (53, 54), as well as deletion of the retinoblastoma and p53 genes (54, 55).

Probes for interphase analysis should give strong, localized signals. For this reason, chromosome-specific centromeric probes are ideal. These give signal that is tightly localized in both the metaphase chromosomes and interphase chromatin such that enumeration of signals can be carried out rapidly and accurately. However, centromeric probes are of limited use for the detection of structural abnormalities. Whole chromosome painting probes are also not suitable for interphase analysis, as the chromosomal domains are often highly extended, making interpretation difficult.

The use of differentially labelled probes flanking specific translocation breakpoints detected with different fluorochromes provides a very specific way of detecting translocations in individual cells. Dual colour FISH with specific loci probes can also be employed for the detection of inversions (56) and amplification (47, 57) in interphase nuclei. This type of analysis may also be linked to the cell morphology or immunophenotype (58). Applications for this type of analysis include the detection of residual leukaemic cells after treatment, and the association of certain abormalities with resistance to treatment (58, 59).

High hybridization efficiencies (85–95%) are required for interphase FISH analysis, particularly for the identification of deletions. Cosmids, or a pool of clones from the region of interest are most suitable: YAC probes often give extended signals in interphase. In the case of deletions, co-hybridization of the test probe with a control probe of similar type will act as an internal control for the hybridization. The cut-off level must be determined for each probe under your own experimental conditions. Careful design of the controls as well as evaluation of the data is extremely important, especially if these procedures are to be used for the detection of minor cell populations.

4.5.1 Combined immunophenotyping and FISH

This technique enables the simultaneous identification of cell type (by immunostaining with specific antibodies) and of chromosome abnormalities (by FISH with specific probes). The correlation of cell type and chromosomal abnormality in individual cells can provide important information about cell lineage involvement in leukaemia and the developmental level of the target cell for clonal abnormalities. This method relies on the ability of the reaction product of the alkaline phosphatase–anti-alkaline phosphatase (APAAP) immunophenotyping method to remain throughout the subsequent hybridization procedures, and to produce red fluorescence that can be viewed at the same time as the fluorescent FISH signal. This approach has been used to study lineage involvement in myeloid disorders (60, 61), and the lineage involvement of trisomy 12 in CLL (62).

4.6 Comparative mapping

Multicolour FISH techniques now provide an exciting prospect for the analysis of evolutionary relationships by comparative cytogenetics (63, 64). This has recently been demonstrated by hybridization of human chromosome painting probes and spectral karyotyping to analyse a highly rearranged gibbon karyotype (6). This approach can also be used to determine human–rodent syntenic regions by hybridizing species-specific paints (or region-specific paints) to human chromosomes, producing a multicoloured 'bar code' (65).

4.7 Gene expression, structure, and function

FISH provides an ideal tool for the examination of the relationship between three-dimensional nuclear structure and function. This has been used to study the organization of RNA during DNA transcription, splicing, and transport (66–68). Interphase FISH also provides a way of analysing the replication status of individual genes, and also to define the boundaries of replication time zones (69, 70). A gene in G1 or early S appears as a single fluorescent signal, whereas a doublet signal is observed in G2, representing the hybridization sites on both chromatids. This assay has been used to demonstrate that the non-transcribed X-linked genes *FMR1* and *XIST* are late replicating (71, 72).

The correlation between chromosomal banding patterns and the distribution of GC isochores in the human genome has been demonstrated by FISH using CpG island DNA (isolated after methylation-sensitive restriction enzyme digest and size fractionation) as a probe (73). This demonstrated that the regions of highest gene density are the early replicating, G-band negative chromosomal regions, and that the highest concentration of genes was clustered in a subset of these (T-bands). The application of FISH to meiotic cells opens up a previously difficult area of research, and promises to clarify many aspects of meiotic chromosome behaviour. This has provided insight into chromosome pairing (74, 75), and the behaviour of human reciprocal translocation chromosomes in heterozygote males (76).

The attraction of FISH for many investigations is the speed, sensitivity, and specificity with which it can be applied to large numbers of cells. Due to advances in both probe labelling and fluorescence detection, both the sensitivity and resolution of FISH mapping techniques have increased, and these techniques now play an integral part in mapping strategies. The last obstacles to multicolour identification of all human chromosomes in a single chromosome painting experiment have now been overcome, opening up the possibility for the analysis of complex karyotypes, and comparative analysis of different species.

References

1. Pinkel, D., Straume, T., and Gray, J. W. (1986). *Proc. Natl. Acad. Sci. USA*, **83**, 2934.
2. van Ommen, G. J. B., Breuning, M. H., and Raap, A. K. (1995). *Curr. Opin. Genet. Dev.*, **5**, 304.
3. Kerstens, H. M., Poddighe, P. J., and Hanslaar, A. G. J. M. (1995). *J. Histochem. Cytochem.*, **43**, 347.
4. Raap, A. K., van de Corput, M. P. C., Vervenne, R. A. W., van Gijlswijk, R. P. M., Tanke, H. J., and Wiegant, J. (1995). *Hum. Mol. Genet.*, **4**, 529.
5. Speicher, M. R., Ballard, S. G., and Ward, D. C. (1996). *Nature Genet.*, **12**, 368.
6. Schrock, E., du Manior, S., Veldman, T., Schoell, B., Wienberg, J., Ferguson-Smith, M. A., *et al.* (1996). *Science*, **273**, 494.
7. Buckle, V. J. and Kearney, L. (1994). *Curr. Opin. Genet. Dev.*, **4**, 374.
8. Lawrence, J. B., Singer, R. H., and McNeil, J. A. (1990). *Science*, **249**, 928.
9. Wiegant, J., Kalle, W., Mullenders, L., Brookes, S., Hoovers, J. M. N., Dauwerse, J. G., *et al.* (1992). *Hum. Mol. Genet.*, **1**, 587.

10. Heiskanen, M., Karhu, R., Hellsten, E., Peltonen, L., Kallioniemi, O. P., and Palotie, A. (1994). *BioTechniques*, **17**, 928.
11. Bensimon, A., Simon, A., Chiffaudel, A., Croquette, V., Helsot, F., and Bensimon, D. (1994). *Science*, **265**, 2096.
12. Weier, H.-U. G., Wang, M., Mullikin, J. C., Zhu, Y., Chang, J.-F., Greulich, K. M., *et al.* (1995). *Hum. Mol. Genet.*, **4**, 1903.
13. Collins, C., Kuo, W. L., Segraves, R., Fuscoe, J., Pinkel, D., and Gray, J. W. (1991). *Genomics*, **11**, 997.
14. Vooijs, M., Yu, L.-C., Tkachuk, D., Pinkel, D., Johnson, D., and Gray, J. W. (1993). *Am. J. Hum. Genet.*, **52**, 586.
15. Pinkel, D., Landegent, J., Collins, C., Fuscoe, J., Segraves, R., Lucas, J., *et al.* (1988). *Proc. Natl. Acad. Sci. USA*, **85**, 9138.
16. Ausubel, F. M., Brent, R., Kingston, R. E., Moore, D. D., Seidman, J. G., Smith, J. A., and Struhl, K. (ed.) (1993). *Current protocols in molecular biology*, 1.13.3, suppl 10. John Wiley & Sons, Inc.
17. Lengauer, C., Green, E. D., and Cremer, T. (1992). *Genomics*, **13**, 826.
18. Heng, H. H. Q. and Tsui, L.-C. (1994). In *In situ hybridization protocols, methods in molecular biology* (ed. K. H. Choo), Vol. 33, p. 35. Humana Press, New Jersey.
19. Lichter, P., Cremer, T., Borden, J., Manuelidis, L., and Ward, D. C. (1988). *Hum. Genet.*, **80**, 224.
20. Cremer, T., Lichter, P., Bordon, J., Ward, D. C., and Manuelidis, L. (1988). *Hum. Genet.*, **80**, 235.
21. Dauwerse, J. G., Wiegant, J., Raap, A. K., Breuning, M. H., and van Ommen, G. J. B. (1992). *Hum. Mol. Genet.*, **1**, 593.
22. Wiegant, J., Wiesmeijer, C. C., Hoovers, J. M. N., Schuuring, E., d'Azzo, A., Vrolijk, J., *et al.* (1993). *Cytogenet. Cell Genet.*, **63**, 73.
23. Haaf, T. and Ward, D. C. (1994). *Hum. Mol. Genet.*, **3**, 697.
24. Laan, M., Kallioniemi, O. P., Hellsten, E., Alitalo, K., Peltonen, L., and Palotie, A. (1995). *Genome Res.*, **5**, 13.
25. Lichter, P., Tang, C.-j. C., Call, K., Hermanson, G., Evans, G. A., Housman, D., *et al.* (1990). *Science*, **247**, 64.
26. ISCN (1995). *An international system for human cytogenetic nomenclature* (ed. F. Mitelman). S. Karger, Basel.
27. Albertson, D. G., Sherrington, P., and Vaudin, M. (1991). *Genomics*, **10**, 143.
28. Fan, Y.-S., Davies, L. M., and Shows, T. B. (1990). *Proc. Natl. Acad. Sci. USA*, **87**, 6223.
29. Baldini, A. and Ward, D. C. (1991). *Genomics*, **9**, 770.
30. Trask, B., Pinkel, D., and van den Engh, G. (1989). *Genomics*, **5**, 710.
31. Florijn, R. J., Bonden, L. A. J., Vrolijk, H., Wiegant, J., Vaandrager, J.-W., Baas, F., *et al.* (1995). *Hum. Mol. Genet.*, **4**, 831.
32. Tocharoentanaphol, C., Cremer, M., Schröck, E., Blonden, L., Kilian, K., Cremer, T., *et al.* (1994). *Hum. Genet.*, **93**, 229.
33. Parra, I. and Windle, B. (1993). *Nature Genet.*, **5**, 17.
34. Florijn, R. J., van de Rijke, F. M., Vrolijk, H., Blonden, L. A. J., Hofker, M. H., den Dunnen, J. T., *et al.* (1996). *Genomics*, **38**, 277.
35. Heiskanen, M., Kallioniemi, P., and Palotie, A. (1996). *Gen. Anal. Biomol. Eng.*, **12**, 179.
36. Vaandrager, J.-W., Schuuring, E., Zwikstra, E., de Boer, C. J., Kleiverda, K. K., van Krieken, J. H. J. M., *et al.* (1996). *Blood*, **88**, 1177.

37. Telenius, H., Pelmear, A. H., Tunnacliffe, A., Carter, N. P., Behmel, A., Ferguson-Smith, M. A., *et al.* (1992). *Genes Chromosomes Cancer*, **4**, 257.
38. Carter, N. P., Ferguson-Smith, M. A., Perryman, M. T., Telenius, H., Pelmear, A. H., Leversha, M. A., *et al.* (1992). *J. Med. Genet.*, **29**, 299.
39. Rack, K. A., Harris, P. C., MacCarthy, A. M., Boone, R., Raynham, H., McKinley, M., *et al.* (1993). *Am. J. Hum. Genet.*, **52**, 987.
40. Flint, J., Wilkie, A. O. M., Buckle, V. J., Holland, A. J., and McDermid, H. E. (1995). *Nature Genet.*, **9**, 132.
41. Meltzer, P. S., Guan, X.-Y., Burgess, A., and Trent, J. M. (1992). *Nature Genet.*, **1**, 24.
42. Guan, X.-Y., Meltzer, P. S., Cao, J., and Trent, J. M. (1992). *Genomics*, **14**, 680.
43. Su, Y. A., Trent, J. M., Guan, X.-Y., and Meltzer, P. S. (1994). *Proc. Natl. Acad. Sci. USA*, **91**, 9121.
44. Rubstov, N., Senger, G., Kuzcera, H., Neumann, A., Kelbova, C., Junker, K., *et al.* (1996). *Hum. Genet.*, **97**, 705.
45. Nederlof, P. M., van der Flier S., Wiegant, J., Raap, A. K., Tanke, H. J., Ploem, J. S. and van, d. P. M. (1990). *Cytometry*, **11**, 126.
46. Ried, T., Baldini, A., Rand, T. C., and Ward, D. C. (1992). *Proc. Natl. Acad. Sci. USA*, **89**, 1388.
47. Kallionemi, A., Kallionemi, O. P., Sudar, D., Rutovitz, D., Gray, J. W., Waldman, F., *et al.* (1992). *Science*, **258**, 818.
48. du Manoir, S., Speicher, M. R., Joos, S., Schrock, E., Popp, S., Dohner, H., *et al.* (1993). *Hum. Genet.*, **90**, 590.
49. Speicher, M. R., du Manior, S., Schrock, E., Holtgreve-Grez, H., Schoell, B., Lengauer, C., *et al.* (1993). *Hum. Mol. Genet.*, **2**, 1907.
50. Giollant, M., Bertrand, S., Verrelle, P., Tchirkov, A., du Manior, S., Ried, T., *et al.* (1996). *Hum. Genet.*, **98**, 265.
51. Poddighe, P. J., Van Der Lely, N., Vooijs, P., De Witte, T., and Ramaekers, F. C. S. (1993). *Exp. Hematol.*, **21**, 859.
52. Bentz, M., Schroder, M., Herz, M., Stilgenbauer, S., Lichter, P., and Dohner, H. (1993). *Leukemia*, **7**, 752.
53. Anastasi, J., Le Beau, M. M., Vardiman, J. W., Fernald, A. A., Larson, R. A., and Rowley, J. D. (1992). *Blood*, **79**, 1796.
54. Döhner, H., Fischer, K., Gabot, G. P., Hansen, K., Pilz, T., Diehl, D., *et al.* (1993). *Blood*, **82**, 2236.
55. Stilgenbauer, S., Dohner, H., Bulgay-Morschel, M., Weitz, S., Bentz, M., and Lichter, P. (1993). *Blood*, **81**, 2118.
56. Dauwerse, J. G., Kievits, T., Beverstock, G. C., van der Keur, D., Smit, E., Wessels, H. W., *et al.* (1990). *Cytogenet. Cell Genet.*, **53**, 126.
57. Kallioniemi, A., Kallioniemi, O. P., Waldman, F. M., Chen, L. C., Yu, L. C., Fung, T. K., *et al.* (1992). *Cytogenet. Cell Genet.*, **60**, 190.
58. Anastasi, J., Feng, J., Le Beau, M. M., Larson, R. A., Rowley, J. D., and Vardiman, J. W. (1993). *Blood*, **81**, 1580.
59. Cano, I., Martinez, J., Quevedo, E., Pinilla, J., Martin-Recio, A., Rodriguez, A., *et al.* (1996). *Cancer Genet. Cytogenet.*, **90**, 118.
60. Kibbelaar, R. E., van Kamp, H., Dreef, E. J., de Groot Swings, G., Kluin Nelemans, J. C., Beverstock, G. C., *et al.* (1992). *Blood*, **79**, 1823.
61. Price, C. M., Kanfer, E. J., Colman, S. M., Westwood, N., Barret, A. J., and Greaves, M. F. (1992). *Blood*, **80**, 1033.

62. Garcia-Marco, J., Jones, D., Colman, S., Matutes, E., Oscier, D., Catovsky, D., *et al.* (1995). *Blood*, **86**, 344a.
63. Scherthan, H., Cremer, T., Arnason, U., Weier, H. U., Lima-de-Feria, A., and Frönicke, L. (1994). *Nature Genet.*, **6**, 342.
64. Richard, F., Lombard, M., and Dutrillaux, B. (1996). *Genomics*, **36**, 417.
65. Muller, U., Rocchi, M., Ferguson-Smith, M. A., and Wienberg, J. (1997). *Hum. Genet.*, **100**, 271.
66. Xing, Y., Johnson, C. V., Dobner, P. R., and Lawrence, J. B. (1993). *Science*, **259**, 1326.
67. Carter, K. C., Bowman, D., Carrington, W., Fogarty, K., McNeil, J. A., Fay, F. S., *et al.* (1993). *Science*, **259**, 1330.
68. Dirks, R. W., Daniël, K. C., and Raap, A. K. (1995). *J. Cell Sci.*, **108**, 2565.
69. Selig, S., Okumura, K., Ward, D. C., and Cedar, H. (1992). *EMBO J.*, **11**, 1217.
70. Kitsberg, D., Selig, S., Brandeis, M., Simon, I., Keshet, I., Driscoll, D. J., *et al.* (1993). *Nature*, **364**, 459.
71. Torchia, B. S., Call, L. M., and Migeon, B. R. (1994). *Am. J. Hum. Genet.*, **55**, 96.
72. Hansen, R. S., Canfield, T. K., and Gartler, S. M. (1995). *Hum. Mol. Genet.*, **4**, 813.
73. Craig, J. M. and Bickmore, W. A. (1994). *Nature Genet.*, **7**, 376.
74. Cheng, E. Y. and Gartler, S. M. (1994). *Hum. Genet.*, **94**, 389.
75. Armstrong, S. J., Kirkham, A. J., and Hulten, M. A. (1994). *Chromosome Res.*, **2**, 445.
76. Rousseaux, S., Chevret, E., Monteil, M., Cozzi, J., Pelletier, R., Devillard, F., *et al.* (1995). *Cytogenet. Cell Genet.*, **71**, 240.

8

Gene expression databases

DUNCAN DAVIDSON, RICHARD BALDOCK, JONATHAN BARD,
MATTHEW KAUFMAN, JOEL E. RICHARDSON, JANAN T. EPPIG,
and MARTIN RINGWALD

1. Introduction

The rapid accumulation of large amounts of information from *in situ* hybridization experiments, together with the spatial and temporal complexity of this kind of data, presents serious difficulties for conventional publishing methods and for the individual researcher who wants to collate and analyse other people's results in relation to his own studies. The solution to these problems is to construct a database of gene expression and several are now being built; most apply to embryos, but in the future it is likely that they will be extended to include the adult. This chapter aims to give a practical overview of the use of gene expression databases in the study of animal development, with particular reference to data from *in situ* hybridization experiments.

One problem encountered in writing about gene expression databases is that they are only now being developed; few are operational at the time of writing. We have not, therefore, attempted to guide the reader through the details of using any particular database. Indeed, each will have its own, periodically updated, detailed instructions at its World Wide Web (WWW) site. Instead, we have tried to give an overview of the field and, in particular, to help the reader appreciate the practical considerations that determine the successful use of any gene expression database. Our examples are mainly drawn from the mouse gene expression information resource (MGEIR).

2. The database approach: combining information in a common format

The essence of the database approach is to place data in a standard format so that it can be stored in retrievable form, collated, and searched. By storing different kinds of data in the same format, it becomes possible to use a complex palette of information to build a picture of gene function in the organism. Thus, for example, combining data from RNA *in situ* hybridization, immunohistochemistry, and histochemistry with data from homogenized material (blots, gels, gridded arrays, etc.) provides comprehensive information about expressed sequences. This approach can, in principle, be extended to an ever widening range of data, some of which may be in different databases on different sites. Using gene name or probe sequence as the common denominator, expression can be related to the genetic and physical location of the gene, its sequence, the phenotypes of mutants, etc. Using the names of those parts of the embryo where the gene is expressed, expression can be linked to information about cellular, physiological, and developmental processes that occur at these locations.

2.1 Descriptions of embryos: information formats for developmental biology

The central concept of a standard format for gene expression data brings into focus certain limitations in our current description of the developing embryo. There are two ways to describe the embryo, words or pictures, and both present problems:

(a) Text is, of course, an important form of knowledge representation, integration, and interrogation which potentially provides links across a very wide range of data. The basic requirement for text description is a standard nomenclature that describes the parts of the embryo, their relationships, and the stages when each part is present. Perhaps surprisingly, for most model organisms this is only now being achieved, but the results will allow gene expression domains to be recorded as standard text which can easily be searched. One problem with the text-based approach is that gene expression domains do not necessarily relate, in time and space, to those divisions of the embryo that have been recognized and named by classical anatomy.

(b) The solution to this problem is to use a graphical, or image mapping approach where the reference format is a set of 2D or 3D images of embryos at successive stages. By mapping the results from different gene expression experiments onto these reference images, it becomes possible to compare different expression patterns within the same spatio–temporal framework and to search the data on the basis of spatial queries that are independent of any anatomical naming system. 3D images have the advantage that they can, in principle, be digitally resectioned, thus making it possible to enter data from experimental embryos that may have been sectioned in any plane. (The image mapping approach is distinct from storing an archive of *independent* images which, while recording original data, does not allow spatial searches.) The graphical approach has two additional advantages. First, unlike the text-based approach, it is not necessary to have identified parts of the embryo by name, a feature of considerable value to those with limited anatomical training (though the entry of data in relation to an image may, if carried out by inexperienced workers, require care if it is not to result in major errors). Secondly, because anatomical structures and expression domains have complex shapes, certain significant spatial relationships within the data may only be perceived in 3D graphical form. A disadvantage of the graphical approach is that each reference image can precisely represent only a single time point in development. In contrast, the text-based approach, by listing all the anatomical components to be found during each successive stage, can describe gene expression at any time in development and is more flexible in this respect. The main issue remains, however, to use the optimum number of reference embryos *or* standard stages as the framework in which to collate information from a large number of independent experiments.

Databases that combine text and graphical descriptions may successfully eliminate the disadvantages of both methods. To achieve this, the two types of description must be linked. This is being done, for both mouse and *Drosophila*, by 'painting' the structures named in the text anatomy onto the reference images.

2.2 Molecular anatomy: a new description of the embryo

The nomenclature of classical anatomy is our '*external*' description of the embryo, while images of the expression domains of different genes, brought together in a single space–time framework, present a picture of the embryo's '*internal*' description (1). This 'molecular anatomy' can, itself, be used to describe the embryo. Invariant, well-defined molecular patterns can be used as spatial markers in much the same way as are recognizable anatomical structures. Similarly, key molecular events during tissue development or cell differentiation can be used as temporal reference points for the expression of other genes. In either case, the expression of a test gene would be compared, in the same experiment, with that of one or more reference genes. One can foresee that our classical morphological description of development will be refined and deepened by this complementary molecular anatomy.

2.3 Using gene expression data to build molecular pathways

A major use of *in situ* gene expression data is to search for gene products with overlapping or complementary distributions in order to build a picture of potential molecular interactions and regulatory pathways. Different developmental systems often make use of common pathways. Consequently, even those with expert knowledge of one system will need to scan data from other systems for relevant information and it seems inevitable that the database approach will provide the only way to do this. There is an important caveat, however: because information in the database has necessarily been transformed into a standard format (textual or graphical), the resolution of the data may not be sufficient for exact comparisons of pattern. In this light, the database is best regarded, not as a repository of truth, but as an 'hypothesis generator', enabling possible relationships to be perceived which must then be tested directly by experiment.

2.4 Linking pathways with processes at the cell and tissue levels

Gene expression data can be related to the cellular and tissue processes that occur during normal morphological development (including, for example, information about cell proliferation, apoptosis, cell migration, and cell shape change) by representing this additional information within the same framework. It may then be possible to make hypotheses about the involvement of molecular pathways in the mechanisms of morphogenesis and differentiation. In particular, those databases that include lineage information will help uncover links between the expression of one gene and that of another at an earlier stage of development. In addition, gene expression databases that are embedded in larger information systems, thus allowing expression to be related to genome sequence and mutant phenotype, will provide increasingly powerful tools to investigate gene function as more sequence information and more mutants become available for each of the model organisms studied by developmental biologists.

2.5 Problems with the database approach

There is, inevitably, some labour in entering information into any database and, for those systems that rely on the efforts of researchers, it will be crucial that the entry of new data is simple and provides tangible benefits. Some of these benefits will be immediate: for example, data entry may help organize laboratory results or help interpret experiments by facilitating the identification of anatomical structures. Once entered into the database, the expression pattern of the test gene can be analysed in relation to other gene expression domains. At first, few databases will contain enough information to be useful in this respect, but the entry of retrospective data from the literature, by database curators, may help to address this difficulty. Importantly, gene expression databases will allow the researcher to publicize purely descriptive results which may be difficult to publish in conventional journals: the MGEIR, for example, plans to offer the option of peer review so that results of this type can achieve full publication status.

One especially challenging problem for the database approach is the patchy nature of a great deal of gene expression data. Although the same difficulties affect the interpretation of results published by conventional means, attempts to codify the data explicitly in a systematic format will bring into focus the incomplete nature of the observations (itself a valuable exercise!). For example, when authors have looked at only a few sections from an embryo, or examined whole mount preparations at low resolution, there may be insufficient evidence to define precisely the tissue distribution of the gene product. The solution adopted by many databases is to store data as far as possible in its original form, and to deal with the problem of incomplete coverage later, when the database is searched. For example, examination of selected sections may suggest that a tissue—say, mesoderm—is uniformly labelled. The researcher enters 'expression in mesoderm', but is required, nevertheless, to indicate which regions were examined. Then, when the database is searched for genes expressed in the mesoderm, the search procedure will uncover all possible instances of expression in this tissue, but the results will be presented in such a way that the user sees the extent of the original observations.

Despite these problems, the gene expression database will rapidly become established as a necessary tool for developmental biologists. Indeed, the difficulties should be viewed in proportion to the sheer complexity of the processes that biologists seek to understand. For the present, the emphasis is on relatively simple database functions: in the longer-term the analysis of gene expression data is likely to become increasingly sophisticated.

3. Outline survey of gene expression databases

In this section we describe, in outline, the main databases that store *in situ* gene expression data. WWW addresses are shown in *Table 1*. The information given is up-to-date at January 1997 and any changes will be found at these WWW addresses.

Table 1. Internet addresses of gene expression, and related, databases

Caenorhabditis elegans database	http://www.sanger.ac.uk/ftp site overview.html http://www.sanger.ac.uk/~silvia/http://www.sanger.ac.uk/~sjj/C.elegans_Home.html
Flyview	http://flyview.uni-muenster.de/
Flybrain project	http://flybrain.uni-freiburg.de/
FlyBase	http://flybase.bio.indiana.edu:82/
Zebrafish database	http://zfish.uoregon.edu/
Xenopus molecular marker resource	http://vize222.zo.utexas.edu/
Xenopus database	http://www.dkfz-heidelberg.de/abt0135/hpeng.htm
Mouse gene expression information resource	http://genex.hgu.mrc.ac.uk/ http://www.informatics.jax.org/
Mouse embryo 2D gel protein database	http://siva.cshl.org/mouse.html
TBASE	http://www.gdb.org/Dan/tbase/tbase.html
dbEST	http://www.ncbi.nlm.nih.gov/dbEST/
Kidney development database	http://mbisg2.sbc.man.ac.uk/kidbase/kidhome.html http://www.ana.ed.ac.uk/anatomy/kidbase/kidhome.html
Organ development database	http://www.ana.ed.ac.uk/anatomy/kidbase/orghome.html
Tooth gene expression database	http://honeybee.helsinki.fi/toothexp/toothexp.html
Ion channel database	http://parrot.le.ac.uk/csn/

3.1 *Caenorhabditis*

Gene expression data for the nematode worm, *C. elegans*, is stored in ACeDB, a free, publicly available database which integrates a wide range of information about this organism. ACeDB is designed to contain, in an easily used format, any type of experimental data, in particular genetic and physical mapping data (clones and contigs), and the complete DNA sequence. Data submitted by users is entered into the public database by the curator. Owners of copies of the database can, however, add data for use on their own machines or run their own database containing expression data. The main interface is for UNIX machines and uses an X-windows, mouse-based, click-and-point navigation method which is easy to learn, and the system provides query and table-making functions, bibliography searches, and general search engines. There are numerous display capabilities including configurable genetic map displays, physical map displays, and sequence–feature displays for DNA, showing restriction sites, splice sites, and protein homologies. The gene expression data is in both image (UNIX versions only) and text form. Images are added to a picture library and can be called from within the database and displayed in a separate viewer (XV) on UNIX machines. Text data are submitted as ASCII files via e-mail or ftp that are read into the database in a standard tree-form structure.

Gene expression data can be searched by text string, or accessed through searches on the other types of data, including individual cells, cell groups, sequences, loci, clones, and papers. 87 gene expression patterns (40 documented with images) have been accumulated from 1990 onwards. Though most of this data comes from just two laboratories, steps are being taken to encourage submission by other workers. It will take some time, however, to enter the large amount of information potentially available. ACeDB is available to authorized sites through the WWW. The database administrators at the Sanger Centre, Cambridge, UK release version code for Sun, Solaris, DEC(OSF), and SGI (IRIX) machines and there is a Mac version, MACACE. There are no immediate plans for links with other organisms.

3.2 *Drosophila*

Three databases are being developed to store gene expression data relating to *Drosophila* development.

(a) A database being built at the University of California (2) aims to model the *Drosophila* embryo at the single cell level. The core of this graphical database is a set of 3D models of embryos each built from differential interference contrast (DIC, Nomarski) images of serial, optical section planes taken at 2 μm intervals through embryos that are either unstained or stained with suitable markers. The major tissues visible under DIC optics (epidermis, musculature, CNS, intestinal tract), as well as cells that have been labelled with specific markers, are drawn onto these section images: the intention is that all the major organ systems will be outlined. Users will be able to enter their data into these section images, adding information on labelled cells, gene expression, etc. using the outlined structures as landmarks. The data will then be visualized in the context of the whole 3D model. The system will be available via the WWW and the software currently being used runs on a Macintosh PC. Single section files can be visualized on most desktop Macintosh machines, though smoothly rendered, semi-transparent models will probably require at least 32 Mb RAM.

(b) Flyview, based at the University of Münster, stores gene expression data for enhancer trap lines as both text and original images. This database plans, in the future, to include the expression of cloned genes. Individual researchers are responsible for their own data, but the information is entered by the Flyview team. The database can be browsed on the WWW and searches are by text description using the controlled vocabulary of FlyBase (the *Drosophila* genetic database, see *Table 1*). Data can be submitted via the WWW or by e-mail or ftp, and images can also be submitted as slides or photographic prints. To date there are around 2000 images and text descriptions in the database, documenting data from about 350 enhancer trap lines. About 60% of the records are linked to FlyBase and/or to the Encyclopaedia of *Drosophila*. There are currently no plans for links with other species. The enhancer trap lines are available to the community and the database contains an index of stocks.

(c) In addition to these whole embryo databases, a database produced as part of the Flybrain Project aims to provide information relating to the structure and function of the *Drosophila* nervous system (3). This database is associated with an atlas of the nervous system based on schematic representations, serial sections, and particularly informative images, all of which will relate to a common co-ordinate system. The aim is that the atlas will provide even the novice with an up-to-date and thorough overview of how the nervous system is constructed. This atlas will be used as the context for data (images linked to hypertext) which will potentially include patterns of expression of genes and regulatory elements, specific anatomical structures, genetic variants, electron micrographs, activity mapping data, etc. Data will primarily be from *melanogaster*, but information from other species of *Drosophila*, and even other *Diptera*, may be included for comparison. Submitted data will be subject to peer review. It is planned to link this database to FlyBase. Flybrain is based at the University of Freiburg; ERATO, Tokyo; and the University of Arizona.

3.3 Zebrafish

For the zebrafish, gene expression data will be stored in a database under development at the University of Oregon. Its main purpose is to provide public access to genetic and developmental data from the zebrafish, with gene expression patterns being only one component of the data. The database will be accessible through a graphical interface to the WWW and, while some of the data will be entered by 'authorized users', some will be entered by the data administrator. In order that even incomplete information will be available to the community, unpublished as well as published data will be included; unpublished information will be marked so that users can assess the confidence level of each entry. The database will store both text and image data and it is planned to have a form of spatial searching based on graphical representations of the embryo. The aim is to include an estimated 6000 images, documenting the expression of 150 genes. In addition, the database will have an atlas for staging embryos (about 40 images of whole embryos and 400 images of 10 μm sections), an anatomical atlas of the adult (150 images of 10 μm sections), information on a large number of mutants (including images), as well as genetic information (gene names, markers, etc.), information on probes, antibodies, etc., a bibliography, and lists of people working on the zebrafish. 'Guest' users will be able to run common types of searches directly from the WWW interface; more unusual queries will be run via the data administrator. In the longer-term, it is planned to link this database to those for other organisms, including the MGEIR.

3.4 *Xenopus*

The *Xenopus* molecular marker resource (XMMR), based at the University of Texas, lists genes and markers with descriptions of gene expression in free text and images of whole mount embryos and sections (about 50 genes or markers are listed at end 1996). Entries include references, a note on the availability of probes, etc., and links to sequence information. Searching is by text and the expression of these markers can be displayed by tissue staining pattern or by gene. The XMMR home page links to a wide range of information on *Xenopus*, including standard stages, and lists of libraries and expression plasmids.

Another database, the Xenopus database, containing information on gene expression in *Xenopus* is being constructed at the DKFZ, Heidelberg. This database, which is not linked to the XMMR, will be in ACeDB format and include text and image files. The current version has expression data from 270 cDNAs isolated in a random *in situ* hybridization screen.

3.5 Mouse

3.5.1 The mouse gene expression information resource

The MGEIR will be a publicly accessible database with text, original images, and image mapped data fully integrated with both the mouse genome database (MGD) and the 3D atlas of mouse development (4, 5). Retrospective data will be entered by database curators. New data will be submitted by individual

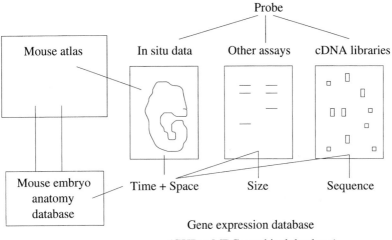

Probe

| Mouse atlas | In situ data | Other assays | cDNA libraries |

| Mouse embryo anatomy database | Time + Space | Size | Sequence |

Gene expression database

(GXD + MRC graphical database)

Figure 1. The organization of the mouse gene expression information resource. The gene expression database will store and integrate data from a variety of assays giving information on sequence, size, and distribution of gene products, information from the different assays being related through details of the sequence or specificity of the probe. The data comprise (in the GXD) text descriptions using the vocabulary of the anatomy database as well as images of original data, and (in the MRC database, which stores only *in situ* data) 3D images of expression domains mapped onto the mouse atlas. The GXD and MRC databases will be closely linked: in addition, the GXD is integrated with the mouse genome database (MGD) which contains genetic, mutant phenotype, and bibliographic data, while the MRC database is integrated with detailed phenotypic information in the mouse atlas.

researchers and checked by editorial staff. Data will include patterns from *in situ* hybridization, histochemistry, immunohistochemistry, and information from assays on homogenized tissues. All the data will be searchable by text queries. Expression patterns that have been spatially mapped onto the mouse atlas will enable spatial searches. The components of the resource are shown in *Figure 1*. The MGEIR comprises the following parts:

(a) The mouse embryo anatomy database.

(b) The mouse atlas.

(c) The GXD gene expression database.

(d) The MRC graphical gene expression database.

i. The mouse embryo anatomy database

This is being built at the University of Edinburgh Anatomy Department, and the MRC Human Genetics Unit. The database will contain the names of morphologically recognizable structures at each of the 26 developmental stages defined by Theiler (6). The names will be organized in a spatial hierarchy

(*Figure 2*) with options for alternative hierarchical schemes. The database will include a thesaurus of alternative names, information on tissue type/architecture, data on tissue derivation and cell lineage (where reliable information exists) to trace the origins and future development of each component, and notes and references on contentious aspects of the nomenclature. The anatomy database will be available on the WWW, with a version accessible via a Java (7) interface.

ii. The mouse atlas

This is being built by the MRC Human Genetics Unit and the University of Edinburgh Anatomy Department. This atlas of embryonic development comprises digital, 3D models of mouse embryos at successive stages from fertilization to stage 22 and will eventually include at least some older embryos. These full grey-level, voxel models will show low magnification views of tissue structure (ranging from 4 µm to 14 µm resolution), and are being reconstructed from images of serial histological sections, generally from the same embryos as illustrated in *The atlas of mouse development* (8). The major named anatomical components are being 'painted' onto the model embryo, transparently overlaying the grey-level images. Custom built software will allow the reconstructed embryo to be resectioned digitally in any orientation and permit each anatomical structure to be tracked through the stack of sections (see *Figure 4*). It will also be possible to view painted anatomical components using commercial software, such as AVS (Advanced Visual Systems. Inc.) (9) (see *Figure 5*) which allows a wide range of manipulations, including stereo 3D visualization using special spectacles. The named structures painted in the reconstructions will link the atlas to the anatomical dictionary, thus making it possible to combine graphical and text descriptions of gene expression domains. In the longer-term, it is anticipated that the mouse atlas will become the focus for a wide range of spatially mapped data relating to mouse embryonic development, bringing this information, in searchable form, into the context of a detailed morphological description of the embryo phenotype. The reconstructed model embryos and painted domains, together with utility software, will be available as a set of CD-ROMs and via ftp.

Figure 2. The mouse embryo anatomy database. Part of the prototype anatomy database is illustrated, showing components of the Theiler stage 14 embryo (E9). The display is shown in 'tree' view to illustrate the hierarchical nature of the database. Part of the tree is 'collapsed' (e.g. extra-embryonic tissue), part is expanded (e.g. embryo...organ system...urogenital system...pronephros); each ticked box can be expanded by clicking on the box to show further components of the hierarchy; clicking on any name calls up a window which allows the user to view data, or enter new information, pertaining to that structure, such as alternative names, lineage, and cell activities. Different parts of the tree can then be viewed at different resolution and the names can be used to describe gene expression domains (see *Figure 3*). The major components of the hierarchy have been painted onto the images of the mouse atlas (see *Figure 5*).

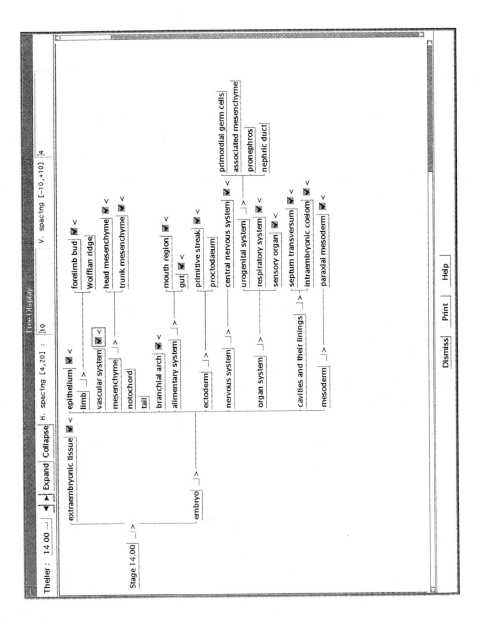

iii. The GXD gene expression database

This is being developed at The Jackson Laboratory, and will store and integrate data from various types of expression assays (RNA *in situ* hybridization, immunohistochemistry, Northern and Western blots, RT-PCR, RNase protection, cDNA arrays, etc.), and thus provide information about what products are produced from a given gene, and about where and when these products are expressed. Expression patterns are described in standardized text using the anatomy database. For *in situ* studies, the textual annotations are complemented with digitized images of original expression data and these images are indexed via the terms from the standard nomenclature. In this way, the expression information shown by the individual images becomes accessible to global text queries.

Expression data will be acquired by annotation from the literature by database editors, and by direct transfer of expression data from laboratories. The gene expression annotator, an electronic data submission system, has been developed to enable comprehensive and standardized description of *in situ* data and electronic transfer of these data from laboratories. The system features selection lists, on-line links to databases, and electronic validation tools to facilitate standardized data annotation and cross-referencing. Digitized images of original *in situ* expression data can be entered into the interface, indexed via the dictionary terms by drag-and-drop procedures, and interactively labelled to provide direct links between the expression domains observed and the standardized anatomical descriptions. Planes of sections are annotated graphically on standardized 2D cartoons. Part of the prototype interface is illustrated in *Figure 3*.

GXD will be fully integrated with the mouse genome database (MGD) to foster a close link to genetic and phenotypic data of mouse strains and mutants. Numerous pointers to other relevant databases will place the gene expression information into the wider biological context.

While the development work on the GXD is ongoing, the database editors identify all newly published research articles documenting data on endo-

---

Figure 3. The GXD gene expression–annotator interface. This part of the prototype interface is used to assemble a text description and original images for submission to the GXD. Other parts of the interface allow interactive entry of information on the gene, embryo studied, experimental method, probe details, and bibliographic data. The controlled vocabulary of the anatomy database is shown on the lower right: items corresponding to gene expression domains are copied by a drag-and-drop mechanism to fields recording experimental results. The top of the figure shows parts of the interface used to enter pictures of the original data. At the top left, an image of a section from the experiment is shown. At the top right, a cartoon of an embryo at the appropriate stage allows the section plane to be recorded. Regions within the heart and neural tube have been interactively labelled on the image to provide a direct link between the observed pattern and the text description.

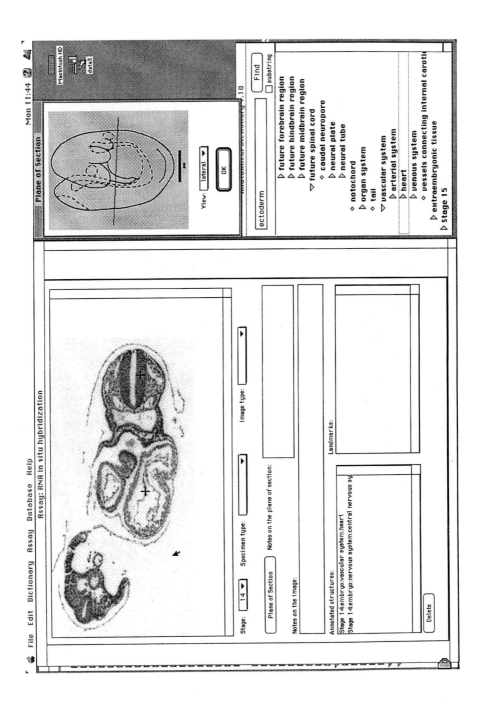

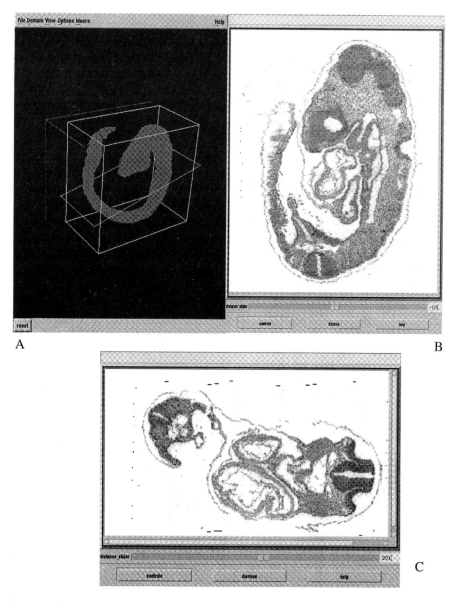

A

B

C

Figure 4. See overleaf for caption.

genous gene expression during mouse development. In a first annotation step, these articles are indexed with regard to authors, journal, genes and embryonic stages analysed, and expression assays used. The GXD index is updated daily, and is available in searchable form via the WWW (*Table 1*, MGEIR, Jackson site).

Figure 4. Finding a plane of section using the mouse atlas. This illustration shows part of the prototype screen painting interface of the mouse atlas. (B) A digital, parasagittal section through a 3D model embryo at Theiler stage 14 (originally reconstructed from images of transverse histological sections). The neural tube and other major organs have been delineated (not shown). (A) The neural tube of the same embryo is shown for purposes of orientation within a box bounding the complete 3D model (the rest of the embryo is not shown). The interface can be used to find a plane of section through this reference embryo that corresponds to the plane in which an equivalent stage, experimental embryo was sectioned for *in situ* hybridization. The user can digitally resection the reference embryo at any angle, using interactive controls, until the reference section shown in (C) matches, as closely as possible, the experimental section as observed under a microscope. The current plane of section is shown automatically in the orientation box in (A): this information can be used, for example, to record the plane of section in the GXD annotator (*Figure 3*). The major components of the embryo have been delineated (not shown here), thus aiding anatomical identification. Having found the appropriate section in the reference image, the same interface can be used with screen painting tools to enter gene expression data from the experiment onto the reference image, and thus to spatially map the domain into the framework of the atlas, allowing the expression of the gene to be compared directly with other, similarly mapped domains. Alternatively, semi-automatic, spatial mapping software can be used to transfer data from an image of the experimental section onto the reference image.

iv. The MRC graphical gene expression database

This is being developed at the MRC Human Genetics Unit and the University of Edinburgh Anatomy Department, and will store graphical gene expression data using 3D images of the mouse atlas as a spatio–temporal framework. Data will be entered locally using custom designed software to map the gene expression domains to the model embryo reconstructions from the atlas CD-ROM (see *Figure 4*). The mapped data will then be transmitted to the database via the WWW. Similarly, graphical queries will be made locally and sent to the remote database; the results will then be returned to be analysed in the context of the atlas. The database is fully integrated with the GXD which contains ancillary data for each entry in the 3D graphical database. Importantly, this integration will enable users to access a wide range of information and to combine text and image methods to enter data and query the database.

The MGEIR is being developed with advice from numerous laboratories in Europe and North America with the aim of developing a user friendly database system for the community at large. Currently, prototypes of the GXD submission interface are being tested by these sites. Although the prototype versions of the GXD interface currently run on Macintosh computers, it is intended that the integrated GXD and MRC gene expression databases will be accessible via the WWW, through platform-independent, textual, and graphical interfaces.

3.5.2 Other gene expression databases for the mouse

In addition to databases that store *in situ* hybridization data, it should be noted that a great deal of gene expression data for mouse (and human) is held

Figure 5. Selected tissues of a Theiler stage 14 embryo from the mouse atlas, displayed as a 3D image. The major tissues of this reference embryo have been delineated and the neural tube, gut, and somites are shown here. Using this type of display, anatomical components and gene expression domains can be viewed in any combination and rotated 3D view. The full, grey-level reconstruction of the model embryo is hidden from view, but a low-resolution section is shown to illustrate how the 3D view can be related to the more detailed, 2D grey-level sections presented by the screen painting interface. The embryo reconstruction is visualized using AVS software (9).

in databases containing information on expressed sequence tags (e.g. dbEST). The mouse embryo 2D gel protein database, at the Cold Spring Harbor Laboratories has information relevant to the expression of protein products, while TBASE, at the John Hopkins University, which stores data on transgenic mice and targeted mutations, includes information on transgene expression in free-text format.

3.6 Human

At least two databases are planned for *in situ* gene expression information from human embryos. One of these, based at the Department of Human

Genetics, University of Newcastle-upon-Tyne, will be publicly accessible via the WWW and will include text and image data. It is planned to make this database compatible with the MGEIR by linking the human anatomical nomenclature system (under construction in the Anatomy Department, University of Edinburgh) with that of the mouse embryo anatomy database. The other, based at the Institute of Child Health, London, is integrated with a graphical atlas of human embryonic development being built at the University of St Andrews. These databases are in the very early stages of development.

3.7 Specialist data

3.7.1 The kidney development database

This publicly accessible database, developed at the University of Edinburgh and the Swiss Federal Institute of Technology, stores gene expression information relating to the developing kidney; similar databases are being constructed for other organs that develop by arborization of an epithelium, for example salivary gland, lung, and mammary gland (see organ development database, *Table 1*). The database contains more than 90% of currently available information, mainly from mouse and *Xenopus*, but zebrafish and a few other animals are included. As well as tables of text data and some original images of gene expression patterns (together about 250 genes), the kidney development database also contains data on more than 30 mutant phenotypes and describes the effects of pharmacological treatments on organ development in culture. Published data is flagged. The data are arranged by tissue type, but not spatially mapped. Queries are supported by a number of possible indexing strategies and more sophisticated searches are planned. The kidney development database can be accessed on the WWW, using Netscape: data can be submitted via the WWW or by e-mail.

3.7.2 Tooth database

The tooth database, based at the University of Helsinki, stores information on gene expression during tooth development in the mouse, rat, and human. The data are in text form and can be searched by text term or by clicking on diagrams of the developing tooth.

4. Using gene expression databases

4.1 Practical considerations in preparing to enter data into a database

Databases differ in their approaches to storing the results of gene expression experiments. Here we discuss, in general terms, the practical considerations to be addressed when preparing data for any such database. Throughout this discussion, the reader should bear in mind the 'golden rule': a simple entry containing only the bare essentials is much better than none at all. *Table 2*,

Table 2. Representative categories of information for a gene expression data entry

Information	Comments relating to MGEIR
Gene name	From MGD; must be accepted gene name
Type of analysis	*In situ* hybridization, histochemistry, cDNA screen, etc.
Method details	Including probe details, etc. that are important for interpretation
Embryo details	Strain, sex, genotype, etc.
Embryo stage	
Tissue examined	From anatomical database
Expression detected/ not detected	
Strength of signal within a domain	e.g. weak, medium, strong
Distribution of signal within a domain	e.g. graded, uniform, striped
Information describing archive image	Section plane marked on diagram provided, magnification, etc.
Spatially mapped images of gene expression	
Ancillary information for mapped images	Method of mapping, illumination, section details, etc.
Results of control experiment	
Link to other, directly comparable entries	Where applicable, links to data from same experiment
Bibliographic reference	Where applicable, may come from MGD
Person responsible	Contact details

which shows the dataset planned for the MGEIR, can be used as a checklist by those planning to use the MGEIR and by researchers working on other organisms to help check which categories of data are stored in their target database.

4.1.1 General considerations

At an early stage in preparing to enter data the user should determine how, and at what resolution, the description of gene expression domains will be stored in the target database. There are several aspects to the question of resolution:

(a) How much detail of the expression pattern can be stored in the database may range from a short description in free text, to a full, 3D image of the pattern. A good gene expression database will be able to describe expression patterns in much more detail, as well as in a more comprehensive and standardized way, than conventional publications. Although most entries in such a database can be expected to make use of these advantages, one exception arises in the case of data from large screens: here, it may be an advantage to submit minimal *in situ* hybridization data, for example on a single section using an uncharacterized, but uniquely identified, probe, providing the data quality is good and the probe is made available. In this way, data which may not be of much interest to the originators would be available to those for whom the additional work in characterizing the gene and its expression pattern is worthwhile. For most databases, the main criterion for accepting data is quality rather than quantity and several databases, including the zebrafish database and the MGEIR, plan to accept data that may be too limited to publish by conventional means.

(b) How the database deals with incomplete data is an important issue (see Section 2.5). If one region, or stage of development, has been the subject of intensive study while others have been given only a cursory examination, this may be recorded so that users can assess the likelihood that small areas, or short periods, of expression have been missed. A related issue is whether, or not, the database records which parts of the embryo, and which stages, were examined but did not show expression, thus distinguishing areas where expression was not detected from those that were not examined: only a few databases, including the MGEIR, explicitly record this data. One important, and especially difficult facet of the problem of incomplete data is presented by bilaterally symmetrical structures. Most records of gene expression patterns, including many published accounts, do not say from which side of the embryo the data comes. (It is often impossible to infer this from a picture of a labelled, transverse section since handedness depends on whether the section is viewed from the rostral or caudal aspect as well as on its orientation.) To allow the possibility that the left and right sides have different expression patterns, database entries should, ideally, state whether expression is observed on the left, right, or both sides of the animal or if the side is not known. Aggregation of data from both sides would then be at the discretion of the user making a query and the source of the raw data would be flagged. This is one of the most difficult problems facing graphical databases and it is not yet clear what solutions will be adopted.

(c) A further aspect is whether or not the database imposes any standardization on the stored data. Most databases (including the MGEIR) are built with the view that stages of development, planes of section, etc. should not necessarily be standardized to conform to the requirements of the database, but that experiments should be designed according to the question being addressed. Where scientific needs permit, however, it is an advantage to use standard stages and standard section planes or views of the embryo so that comparisons between data are easier and more valid. Guidance on standards should be sought from the database WWW site at an early stage in designing the experiment.

4.1.2 Experimental details

Some of the information in *Table 2* describes experimental methods. Simple databases do not store this information, but more sophisticated databases do so in order to meet the basic requirements that the experiment can be properly interpreted and in order to place expression data into the larger biological and analytical context.

A standardized description of the genotype of the embryo is, for example, important if one wishes to compare the expression pattern of a given gene in wild-type animals and in different mutants, or to correlate expression with phenotype in a given mutant. Information describing the genetic strain may be especially important in the future for the discovery of modifiers and the analysis of quantitative traits.

An exact description of the probe used in an *in situ* hybridization experiment is essential to interpret which transcripts are being assayed. Some databases, including the MGEIR, plan to describe the sequence of the probe via detailed cross-references to sequence databases. These annotations enable integration with, and searches of, the rapidly accumulating information contained in sequence databases. The gene expression database can thus automatically check possible cross-hybridization of a probe with newly discovered transcripts, or determine which exons a given cDNA probe will detect, and so evaluate evidence for alternative transcripts.

4.1.3 Temporal information on gene expression

In many animals, including mice, developmental stage varies significantly among embryos of the same post-conception age and even within a litter. Therefore, where possible, embryos should be described using standard staging systems or other generally recognized criteria such as somite number, rather than simply by days post-conception. Most databases either provide, or are linked to, descriptions of standard stages. The MGEIR uses Theiler stages supplemented with subdivisions of the period of gastrulation defined by Downs and Davies (10), and the mouse embryo anatomy database will include simple diagrams of embryos at each stage, a list of structures that characterize each stage, and a conversion table for different staging criteria. The Downs and Davies stages, strictly, apply to mice of the PO and related (Swiss) strains. In the MGEIR, these staging criteria have been adapted and extended to accommodate embryos of the same genetic background as those described in *The atlas of mouse development* (8) (i.e. C57BL × CBA F2 embryos): these embryos, which are representative of laboratory mouse strains commonly used in genetic studies, do not, in all respects, follow the Downs and Davies criteria between primitive-streak and late neural-plate stages (K. Lawson, personal communication).

In addition to the overall changes between successive embryo stages, certain changes in cell morphology or in the expression of molecular markers signify key events during tissue differentiation and thus occur at different times in different parts of the embryo. The expression of a test gene may be compared to these markers, in the same experiment and through successive embryo stages, in order to document its activity in relation to the differentiation process.

4.1.4 Spatial information on gene expression

There are a number of ways in which spatial data can be stored.

1. Spatial co-ordinates of the expression domain can be stored as graphical records. Several databases (including the MGEIR) plan to store spatially mapped data together with such necessary ancillary information as plane of section, number of sections examined, and nominal section thickness. The MGEIR will provide facilities to find the plane of section in a test embryo by finding the most closely matched plane in the equivalent reference embryo in the mouse atlas (see for example, *Figure 4*); this plane can be recorded in the GXD (*Figure 3*). Expression domains can then be mapped onto the reference embryo using tools for screen painting or by semi-automatic image mapping of a digital picture of the data. Image mapping will be done section-by-section, using tie-points to mark corresponding positions in the experimental and reference images. Where the test embryo has adopted a shape different from that of the reference embryo, different angles of section may be needed for different regions. Differences in precise orientation and developmental stage between experimental and reference material will provide a considerable challenge for this approach, but the method has several potential advantages, particularly speed and ease of use.

It is intended that, for the MGEIR, graphical and text interfaces will run side-by-side to permit dual operation. Indeed, both screen painting and image mapping can be aided by text entry. For example, a domain may be input first as text (e.g. 'neural tube') corresponding to an already painted anatomical domain (or to a gene expression domain already in the database): this preliminary domain can then modified by painting to match the observed pattern. The image mapping approach is unlikely to map expression patterns with sufficient accuracy to be used alone and it may be necessary to use text qualifiers (e.g. 'expressed in epithelium, not in mesoderm'), or screen painting, to maximize the match between entered data and observed pattern. Although these methods could, in principle, allow users to enter inaccurate or distorted data, the same is true of conventional methods including straightforward text description: researchers should therefore see these tools simply as a means to enter information easily and with the maximum accuracy.

2. Relation to named anatomical structures can be stored as text records. This is the principal medium which most databases use to record expression. As an example, part of the GXD data entry interface is illustrated in *Figure 3*. Some databases record expression domains in free text; others use standardized nomenclature to ensure consistency and to facilitate the search procedure. Standard descriptions of anatomical components have been devised for *C. elegans* (cell descriptions), *Drosophila* (FlyBase, *Table 1*), and mouse (mouse embryo anatomy database), and a system for human embryos is now accessible. These vocabularies may describe different levels of anatomical

structure, for example, gross morphology, body parts, tissues, cell types, and organelles. In the MGEIR, the hierarchical organization of the nomenclature allows it to be extended and, importantly, allows anatomical structure and gene expression patterns to be described at different levels of resolution (*Figure 2*). Some databases include spatial text qualifiers such as anterior, posterior, dorsal, ventral, etc. as part of a controlled text entry or simply as a comment.

In order to use these standard nomenclatures, the user must first identify named structures in the specimen. Whole mount preparations and histological sections present different problems in this regard. Whole mount preparations, while convenient and widely used in screens, may be too opaque to allow precise determination of which tissues are labelled. Notable exceptions are *C. elegans*, *Drosophila*, and zebrafish embryos where tissues can usually be distinguished using DIC optics. Mouse embryos, however, present a serious problem. Help in identifying tissues may be obtained by viewing the embryos of the mouse atlas where each major internal tissue is named and can be independently displayed within a semi-transparent 3D model that resembles a whole mount preparation, but definitive evidence may require embryos to be studied in section. Histological sections generally give a clearer view of which tissues contain hybridization signal, but the overall form of the expression domain may be difficult to perceive and small areas of expression may be missed that would be apparent in whole mounts. For the mouse, the annotated reconstructions in the mouse atlas viewed in the section plane that best matches the experimental material, will assist in identifying anatomical structures (*Figure 4*). In addition, the 3D reconstruction and image mapping tools provided by the mouse atlas can help the user to perceive the shape of the expression domain (*Figure 5*).

The MGEIR will also include facilities to record the strength of signal (as 'weak, medium, strong') and, it's distribution within the domain ('uniform, striped, graded', etc.).

3. Images of original expression data constitute the ultimate reference for data interpretation. Most databases, therefore, provide access to images of experimental material.

4. Some databases (for example, the MGEIR) may require the submission of sufficient control data to establish the validity of the results. It is not clear, at present, what form this data will take or how it will be recorded.

5. Differences in experimental conditions, including the use of embryos at different stages or sectioned in different planes, critically affect the confidence with which the results of different gene expression experiments can be compared. Therefore, some databases plan to explicitly record which patterns have been compared directly in the same experiment.

4.2 Querying the database

The key to obtaining useful answers to searches is to frame the query with precision. A typical question is 'What genes are co-expressed with gene X during developmental stages A to D?' Queries like this one, containing the most common elements, gene, time, and space, can be made in most databases. The power of the search can be increased, however, by appropriate use of additional information and some databases can search experimental, or bibliographic details, dates, etc. Where expression data is part of a larger database that contains information on DNA sequence, mutants, or cell lineage, this can also be used to build powerful queries.

The resolution with which each of these elements is applied, and the way they are combined are also important in determining the power of the search. For example, suppose one is interested in finding genes that interact during limb development with a known gene, X, whose expression has been characterized, and that genetic evidence suggests that an interacting gene lies within a certain genetic interval. One may search for a candidate gene using the query 'show all genes that are co-expressed with gene X (anywhere in the embryo), **or** which have alleles with mutant phenotypes affecting structures where gene X is expressed, **but exclude** genes known to lie outside the defined genetic interval'. This query is based on the expression pattern of gene X, but makes use of wide ranging information in the database. It is designed to cope with the fact that, although the candidate gene is expected to be expressed in the limb, evidence for this may not be in the database; moreover, the gene that is being sought may be co-expressed with X elsewhere in the embryo and may have alleles, additional to the one being studied, that do not have a limb phenotype. In designing queries of this type, it is important to bear in mind that not all the information in the database will have been entered at the same resolution. Queries on expression in the inner ear might search for all data that could have included expression in the inner ear (including data recorded simply as expression in the ear, or even in the head).

Once the logic of the question has been framed, it is necessary to decide whether the query should be made through the medium of text, diagrams, or graphics.

(a) Text queries can, in principle, provide complex logical combinations and access a wide range of data. Controlled vocabularies, combined with drag-and-drop interface methods can simplify query making and greatly reduce errors. Text query procedures are often very rapid.

(b) Some databases, for example the tooth database, allow the option of searching for genes expressed in particular tissues by clicking within these tissues as represented in simple diagrams. While this method does not distinguish between expression domains that fill different parts of the same

211

tissue, it does present a simple interface that requires no specialist anatomical knowledge.

(c) Graphical query methods are only now being developed for gene expression databases. These will allow searches to find expression domains in relation to an arbitrarily defined region in 3D space and will thus distinguish between expression in different parts of a named anatomical structure. The approach will also allow queries on the basis of relative position, for example, 'Which genes have expression domains that are complementary to the expression of gene X at stages 15 to 18?', or 'Which genes are expressed close to the domain of expression of gene X?' ('close to', being defined by the user in μm).

4.2.1 Following up the return from a query

The results of a search need not be returned in the same medium as used for the query. For example, the answer to a graphical query may be returned as a text list which can be rapidly transmitted across the WWW: the user would then select items for study in a graphical display.

The return from a search may be followed up in a number of ways that make further use of the database. For example, literature references from the database may give additional details of expression patterns and methods or may provide further information on gene function. The database may also contain information on probe details, sources of mutant animals, contact addresses, etc. that will aid the investigation of gene function.

5. The future of gene expression databases

In the present phase of intense data collection, gene expression databases are likely to be used mainly to store, collate, and scan information. Despite the fact that the volume of data is so large as to present serious problems of handling and comprehension, the information is still sketchy in its coverage of developmental processes and the current possibilities for exploring gene function are thus limited. Key players in even small regulatory gene networks are still unknown and we expect that gene expression databases will play an important part in filling out this picture. As more of these players are recognized and their expression patterns characterized, it will become increasingly common to use the databases to frame testable hypotheses regarding genetic interactions in development.

In the next phase, the emphasis is likely to shift from recording and searching gene expression information towards relating the data to our knowledge of other measures of gene function. This knowledge will include genetic and biochemical evidence about interactions between gene products, as well as evidence on the distribution of cellular and morphogenetic processes within the embryo. Linking these data will involve transactions between different databases and will thus depend crucially on establishing compatibility between data used in these different contexts. Gene expression databases are likely to play an important role in establishing standards in this wider field. Cross-referencing information in this way will provide potentially immense power for assembling complex data combinations and anchoring them to genetic organization on one hand, and molecular and morphological anatomy on the other.

5.1 Between-species links

The finding that diverse developmental mechanisms share common regulatory pathways has led to an awareness of the evolutionary dimension of development and, consequently, there is a pressing practical need to integrate information from different model organisms. As the genomes of these organisms are sequenced and more similarities and differences at the genetic level are recognized, it will become increasingly important to compare the functions of their related genes. Although it cannot be taken for granted that expression indicates function, it is clear that cross-species comparisons of gene expression patterns will play an important part in these functional and evolutionary analyses. One way to link related information from different developing systems is to abstract data from different databases and combine them in a single resource aimed at a particular audience. An excellent example of this approach is the ion channel database (*Table 1*). A potentially more flexible approach is to ensure not only that the data are compatible, but that the databases are interoperable. Of the various ways to integrate different databases, the emerging standard for database access and interoperability is the common object request broker architecture (CORBA) (11). If the considerable practical difficulties in this area can be overcome, then the various databases, each designed and maintained by a section of the biomedical community, may be brought together to create a single, distributed database that the individual scientist can trawl in pursuit of his, or her, particular interests.

Acknowledgements

It is a pleasure to thank our colleagues at the MRC Human Genetics Unit (Christophe Dubreuil, Elizabeth Guest, Bill Hill, Jane Quinn, and Margaret Stark), the University of Edinburgh Anatomy Department (Renske Brune), and The Jackson Laboratory (Geoffrey Davis, Laura Trepanier, Dale Begley, and Alex Smith), and Albert Burger and Joe Nadeau for their contributions to the MGEIR, Klaus Schughart for interesting discussions that have been important in developing the ideas presented here, and members of the ESF Network on Gene Expression Databases for helpful discussions. We also thank Silvia Martinelli, Richard Durbin, Volker Hartenstein, Wilfried Janning, Monte Westerfield, Peter Vize, Christof Niehrs, Tom Strachan, and Jamie

Davies, for providing information on their gene expression databases, and Kirstie Lawson for advice on staging mouse embryos.

References

1. Brenner, S. Quoted in Judson, H.F. (1979). *The eighth day of creation.* Jonathan Cape, London.
2. Hartenstein, V., Lee, A., and Toga, A.W. (1995). *Trends Genet.,* **11**, 51.
3. Heisenberg, M. and Kaiser, K. (1995). *Trends Neurosci.,* **18**, 481.
4. Baldock, R.A., Bard, J.B.L., Kaufman, M.H., and Davidson, D. (1992). *BioEssays,* **14**, 501.
5. Ringwald, M., Baldock, R.A., Bard, J.B.L., Kaufman, M.H., Eppig, J.T., Richardson, J.E., Nadeau, J.H., and Davidson, D. (1994). *Science,* **265**, 2033.
6. Theiler, K. (1989). *The house mouse: atlas of embryonic development* (2nd printing). Springer–Verlag, New York.
7. Java soft http://www.javasoft.com/
8. Kaufman, M.H. (1992). *The atlas of mouse development.* Academic Press, London.
9. Advanced Visual Systems, Inc. http://www.avs.com/
10. Downs, K.M. and Davies, T. (1993). *Development,* **118**, 1255.
11. Object Management Group http://www.omg.org/

A1

List of suppliers

This core list of suppliers appears in all books in the Practical Approach series. If there are any relevant suppliers that you would like to add to this list for the book you are working on, please send them with your chapter.

Agar Scientific, 66A Cambridge Road, Stansted CM26 8DA, UK.
American National Can, Neenah, WI 54956, USA.
Amersham
Amersham International plc., Lincoln Place, Green End, Aylesbury, Buckinghamshire HP20 2TP, UK.
Amersham Corporation, 2636 South Clearbrook Drive, Arlington Heights, IL 60005, USA.
Anderman
Anderman and Co. Ltd., 145 London Road, Kingston-Upon-Thames, Surrey KT17 7NH, UK.
Beckman Instruments
Beckman Instruments UK Ltd., Progress Road, Sands Industrial Estate, High Wycombe, Buckinghamshire HP12 4JL, UK.
Beckman Instruments Inc., PO Box 3100, 2500 Harbor Boulevard, Fullerton, CA 92634, USA.
Becton Dickinson
Becton Dickinson and Co., Between Towns Road, Cowley, Oxford OX4 3LY, UK.
Becton Dickinson and Co., 2 Bridgewater Lane, Lincoln Park, NJ 07035, USA.
Bio
Bio 101 Inc., c/o Stratech Scientific Ltd., 61–63 Dudley Street, Luton, Bedfordshire LU2 0HP, UK.
Bio 101 Inc., PO Box 2284, La Jolla, CA 92038–2284, USA.
Biocell Laboratory, Golden Gate, TY Glas Avenue, Cardiff CF4 5DX, UK.
Bio-Rad Laboratories
Bio-Rad Laboratories Ltd., Bio-Rad House, Maylands Avenue, Hemel Hempstead HP2 7TD, UK.
Bio-Rad Laboratories, Division Headquarters, 3300 Regatta Boulevard, Richmond, CA 94804, USA.

Biosynth AG, PO Box 125, 9422 Staad, Switzerland.

Boehringer Mannheim

Boehringer Mannheim UK (Diagnostics and Biochemicals) Ltd., Bell Lane, Lewes, East Sussex BN17 1LG, UK.

Boehringer Mannheim Corporation, Biochemical Products, 9115 Hague Road, PO Box 504 Indianopolis, IN 46250–0414, USA.

Boehringer Mannheim Biochemica, GmbH, Sandhofer Str. 116, Postfach 310120 D-6800 Ma 31, Germany.

British Drug Houses (BDH) Ltd., Poole, Dorset, UK.

CP Laboratories, PO Box 22, Bishop's Stortford CM23 3DX, UK.

DAKO, 22 The Arcade, The Octagon, High Wycome HP11 2HT, UK.

Denley Instruments Ltd., Natts Lane, Billingshurst RH14 9EY, UK.

Difco Laboratories

Difco Laboratories Ltd., PO Box 14B, Central Avenue, West Molesey, Surrey KT8 2SE, UK.

Difco Laboratories, PO Box 331058, Detroit, MI 48232–7058, USA.

Du Pont

Dupont (UK) Ltd. (Industrial Products Division), Wedgwood Way, Stevenage, Hertfordshire SG1 4Q, UK.

Du Pont Co. (Biotechnology Systems Division), PO Box 80024, Wilmington, DE 19880–002, USA.

Enzo Biochem, Inc., 325 Hudson Street, New York, NY 10013, USA.

European Collection of Animal Cell Culture, Division of Biologics, PHLS Centre for Applied Microbiology and Research, Porton Down, Salisbury, Wiltshire SP4 0JG, UK.

Falcon (Falcon is a registered trademark of Becton Dickinson and Co.)

Fisher Scientific Co., 711 Forbest Avenue, Pittsburgh, PA 15219–4785, USA.

Flow Laboratories, Woodcock Hill, Harefield Road, Rickmansworth, Hertfordshire WD3 1PQ, UK.

Fluka

Fluka-Chemie AG, CH-9470, Buchs, Switzerland.

Fluka Chemicals Ltd., The Old Brickyard, New Road, Gillingham, Dorset SP8 4JL, UK.

Gibco BRL

Gibco BRL (Life Technologies Ltd.), Trident House, Renfrew Road, Paisley PA3 4EF, UK.

Gibco BRL (Life Technologies Inc.), 3175 Staler Road, Grand Island, NY 14072–0068, USA.

Arnold R. Horwell, 73 Maygrove Road, West Hampstead, London NW6 2BP, UK.

Hybaid

Hybaid Ltd., 111–113 Waldegrave Road, Teddington, Middlesex TW11 8LL, UK.

Hybaid, National Labnet Corporation, PO Box 841, Woodbridge, NJ 07095, USA.

HyClone Laboratories, 1725 South HyClone Road, Logan, UT 84321, USA.

Ilford Ltd., Town Lane, Mobberley WA16 7JL, UK.

International Biotechnologies Inc., 25 Science Park, New Haven, Connecticut 06535, USA.

Invitrogen Corporation

Invitrogen Corporation, 3985 B Sorrenton Valley Building, San Diego, CA 92121, USA.

Invitrogen Corporation, c/o British Biotechnology Products Ltd., 4–10 The Quadrant, Barton Lane, Abingdon, Oxon OX14 3YS, UK.

Jackson ImmunoResearch

Jackson ImmunoResearch Laboratories Inc., 872 West Baltimore Pike, PO Box 9, West Grove, PA 19390, USA.

Jackson ImmunoResearch, c/o Stratech Scientific Ltd., 61–63 Dudley Street, Luton LU2 0NP, UK.

Kirkegaard and Perry Laboratories

Kirkegaard and Perry Laboratories (KPL), 2 Cessna Court, Gaithersburg, MD 20879-4174, USA.

Kirkegaard and Perry Laboratories, c/o Insight Biotechnology Ltd., PO Box 520, Wembley, London HA9 7YN, UK.

Kodak: Eastman Fine Chemicals, 343 State Street, Rochester, NY, USA.

Raymond A. Lamb, 6 Sunbeam Road, London NW10 6JL, UK.

Life Technologies Inc., 8451 Helgerman Court, Gaithersburg, MN 20877, USA.

Merck

Merck Industries Inc., 5 Skyline Drive, Nawthorne, NY 10532, USA.

Merck, Frankfurter Strasse, 250, Postfach 4119, D-64293, Germany.

Millipore

Millipore (UK) Ltd., The Boulevard, Blackmoor Lane, Watford, Hertfordshire WD1 8YW, UK.

Millipore Corp./Biosearch, PO Box 255, 80 Ashby Road, Bedford, MA 01730, USA.

Molecular Probes Inc., 4849 Pitchford Avenue, Eugene, OR 97402-9144, USA.

National Diagnostics Ltd., Unit 4, Fleet Business Park, Itlings Lane, Hessle, Hull HU13 9LX, UK.

New England Biolabs (NBL)

New England Biolabs (NBL), 32 Tozer Road, Beverley, MA 01915–5510, USA.

New England Biolabs (NBL), c/o CP Labs Ltd., PO Box 22, Bishops Stortford, Hertfordshire CM23 3DH, UK.

Nikon Corporation, Fuji Building, 2–3 Marunouchi 3-chome, Chiyoda-ku, Tokyo, Japan.

Oxoid, Unipath Ltd., Wade Road, Basingstoke RG24 0PW, UK.

Perceptive Scientific International (PSI) Ltd., Halladale, Lakeside, Chester Business Park, Wrexham Road, Chester CH4 9QT, UK.

Perkin-Elmer

Perkin-Elmer Ltd., Maxwell Road, Beaconsfield, Buckinghamshire HP9 1QA, UK.

Perkin Elmer Ltd., Post Office Lane, Beaconsfield, Buckinghamshire HP9 1QA, UK.

Perkin Elmer-Cetus (The Perkin-Elmer Corporation), 761 Main Avenue, Norwalk, CT 0689, USA.

Pharmacia Biotech Europe, Procordia EuroCentre, Rue de la Fuse-e 62, B-1130 Brussels, Belgium.

Pharmacia Biosystems

Pharmacia Biosystems Ltd. (Biotechnology Division), Davy Avenue, Knowlhill, Milton Keynes MK5 8PH, UK.

Pharmacia LKB Biotechnology AB, Björngatan 30, S-75182 Uppsala, Sweden.

Polysciences Europe GmbH, Postfach 11 30, Handelsstr. 3, D-69214 Eppelheim, Germany.

Promega

Promega Ltd., Delta House, Enterprise Road, Chilworth Research Centre, Southampton, UK.

Promega Corporation, 2800 Woods Hollow Road, Madison, WI 53711–5399, USA.

Qiagen

Qiagen Inc., c/o Hybaid, 111–113 Waldegrave Road, Teddington, Middlesex TW11 8LL, UK.

Qiagen Inc., 9259 Eton Avenue, Chatsworth, CA 91311, USA.

Schleicher and Schuell

Schleicher and Schuell Inc., Keene, NH 03431A, USA.

Schleicher and Schuell Inc., D-3354 Dassel, Germany.

Schleicher and Schuell Inc., c/o Andermann and Co. Ltd.

Shandon Scientific Ltd., Chadwick Road, Astmoor, Runcorn, Cheshire WA7 1PR, UK.

Sigma Chemical Company

Sigma Chemical Company (UK), Fancy Road, Poole, Dorset BH17 7NH, UK.

Sigma Chemical Company, 3050 Spruce Street, PO Box 14508, St. Louis, MO 63178–9916, USA.

Sorvall DuPont Company, Biotechnology Division, PO Box 80022, Wilmington, DE 19880–0022, USA.

Sterilin, Bibby Sterilin Ltd., Tilling Drive, Stone ST15 0SA, UK.

Stratagene

Stratagene Ltd., Unit 140, Cambridge Innovation Centre, Milton Road, Cambridge CB4 4FG,UK.

Stratagene Inc., 11011 North Torrey Pines Road, La Jolla, CA 92037, USA.

Taab Laboratories Equipment Ltd., 3 Minerva House, Calleva Park, Aldermaston, RG7 8NA, UK.

Techne (Cambridge) Ltd., Duxford, Cambridge CB2 4PZ, UK.

United States Biochemical, PO Box 22400, Cleveland, OH 44122, USA.

Vector Laboratories Ltd., 16 Wulfric Square, Peterborough PE3 8RF, UK.

Wellcome Reagents, Langley Court, Beckenham, Kent BR3 3BS, UK.

Worthington Biochemical Corporation, Freehold, NJ 07728, USA.

Index

acetylation 12, 42, 49, 76–7
acrylamide, stock solution 30
alkaline phosphatase
  conjugates, *see* hapten-labelled probes
  endogenous, inhibition of 16, 50, 97
  inactivation 16, 19–20, 113–16, 118
3-aminopropylethoxysilane (TESPA),
    see subbing slides
antibodies, suppliers
  anti-antibody 108, 172
  anti-hapten 172
  AP-conjugated anti-hapten 50, 96, 109
  fluorescent 172
  gold-conjugated 137
  HRP-conjugated 108–9
antifading agent 53, 119
AP inactivation buffer 108
artefacts 52, 56; *see also* controls, non-specific
    background, troubleshooting
autoradiography
  developing 59–60, 81
  with liquid emulsion 55–60, 79–81
  removal of background 18
  theory 9, 14–15, 34–5
  with X-ray film 48, 55
  see also quantification of signal
autoradiography film, suppliers 48

base pairing, *see* hybridization theory
BCIP, stock solution 50, 96, 108
biotin, problems with 20, 38, 107, 125, 132;
    see also hapten labelled probes, relative
    merits of
BLAST searches 27–8
blocking powder, stock solution (MABTB)
  97, 108
blocking solution 108
bromophenol blue, stock solution 30

cell culture, for *in situ* hybridization 40–1
CHAPS 95–6, 108
chrome alum gelatin, stock solution 73
chromogenic substrates
  BM Purple 96
  ELF™ 108, 117–18
  Fast Red 108, 110, 112–16
  Magenta-phos 108
  NBT/BCIP 12, 50–1, 96–9, 109, 111,
    113–16
  TrueBlue™ 109–10, 112–13
  Vector Red 109, 113

chromosome
  banding 177
  painting 180–2
chromosome spreads
comparative genomic hybridization 182–3
competitive *in situ* suppression 168–171
concentration of probe, *see* probe
    concentration
controls 18–20, 53–5, 114, 141–2, 181, 184;
    see also troubleshooting
cross-hybridization 6, 18, 89, 208
cross-species hybridization 5–6
cryoprotection solution 42
cryostat, *see* sectioning

DAB staining solution 108
DAPI, supplier 172
dark-field illumination 15, 60–1, 63, 104
databases, gene expression
  Caenorhabditis 194
  Drosophila 194–5
  entering data 205–210
  human 204–5
  Internet addresses 193
  kidney 205
  mouse 196–204
  problems 192–3
  querying 211–12
  tooth 205
  Xenopus 196
  zebrafish 195–6
denaturation of target DNA 132, 134, 139–40,
  169
Denhart's solution, stock solution 65, 108
DEPC treatment, *see* ribonuclease
developing of autoradiographic signal,
    see autoradiography
developing solution for liquid emulsion 59,
  81
dextran sulfate, supplier 45
differential interference contrast illumination
  20, 104
DNA, denatured herring sperm 65
DNA–DNA hybrids 12–13
DNA probes
  use for FISH 168–71
  nick translation, labelling method 165–7
  relative merits 7, 8, 132
  synthesis by PCR 7, 163–4
  see also oligonucleotide probes
DNase, stock solution 141, 166